CW01198036

61
Pm
62
Sm
Samarium
150.36
63
Eu
Europium
151.964
64
Gd
Gadolinium
109
Mt
Meitnerium
(266)
110
(269)
HO
CH3
CH3COOH
base
H2O
CO2
hydrolysis

3

Letts

GCSE (9–1)
Chemistry
Paper 1H

H

Higher tier
Time: 1 hour 45 minutes

Materials

For this paper you must have:
- a ruler
- a calculator
- the periodic table.

Instructions

- Answer **all** questions in the spaces provided.
- Do all rough work in this book. Cross through any work you do not want to be marked.

Information

- There are 100 marks available on this paper.
- The marks for questions are shown in brackets.
- You are expected to use a calculator where appropriate.
- You are reminded of the need for good English and clear presentation in your answers.
- When answering questions 04 and 09 you need to make sure that your answer:
  - is clear, logical, sensibly structured
  - fully meets the requirements of the question
  - shows that each separate point or step supports the overall answer.

Advice

In all calculations, show clearly how you work out your answer.

Name: _____

0 1 This question is about the alkali metals and their compounds.

Table 1 shows the melting points of some alkali metals.

Table 1

Element	Melting point in °C
Lithium	181
Sodium	98
Potassium	63
Rubidium	39

0 1 . 1 Why does rubidium have a lower melting point than potassium? **[1 mark]**

Tick **one** box.

Rubidium is ionic and contains ionic bonds and potassium is covalent and contains covalent bonds. ☐

Rubidium is ionic and potassium is metallic. ☐

The forces between the rubidium ions and the delocalised electrons are weaker than the forces between the potassium ions and the delocalised electrons. ☐

The forces between potassium molecules are stronger than the forces between rubidium molecules. ☐

0 1 . 2 A redox reaction takes place when Group 1 metals react with chlorine.

Balance the equation for the reaction between sodium and chlorine. Include state symbols.

[2 marks]

_____ Na _____ + Cl₂ _____ → _____ NaCl _____

0 1 . 3 Why does solid sodium chloride **not** conduct electricity? **[1 mark]**

Tick **one** box.

It contains a non-metal. ☐

The ions cannot move. ☐

It contains a metal. ☐

It contains ions. ☐

0 2 This question is about the reactions of the halogens.

Table 2 shows the colour of the halogens in aqueous solution.

Table 2

Element	Colour of the aqueous solution
Fluorine	Colourless
Chlorine	Green
Bromine	Orange
Iodine	Brown

0 2 . 1 A redox reaction takes place when bromine is added to potassium iodide.

Complete the word equation that sums up the reaction that takes place. **[1 mark]**

bromine + potassium iodide → _____ + _____

0 2 . 2 Look at **Table 2**. What is the colour of the final solution? **[1 mark]**

_____

0 2 . 3 Complete the ionic equation for the reaction between bromine and potassium iodide.

$Br_2 + 2I^- \rightarrow$ _____ + _____ **[2 marks]**

0 2 . 4 Electrolysis can be used to separate the components in a solution of aqueous potassium bromide.

Identify the products of the electrolysis of aqueous potassium bromide. **[1 mark]**

Tick **one** box.

Product at cathode	Product at anode	
hydrogen	oxygen	☐
hydrogen	bromine	☐
potassium	oxygen	☐
potassium	bromine	☐

0 3 An atom of phosphorus has the symbol $^{31}_{15}P$.

0 3 . 1 Give the number of protons, neutrons and electrons in this atom of phosphorus. **[3 marks]**

Number of protons = _____

Number of neutrons = _____

Number of electrons = _____

0 3 . 2 Why is phosphorus in Group 5 of the periodic table? **[1 mark]**

_____

0 4 **Figure 1** shows an atom of the Group 2 element magnesium and an atom of the Group 7 element chlorine. Only the outer shell of electrons is shown.

Figure 1

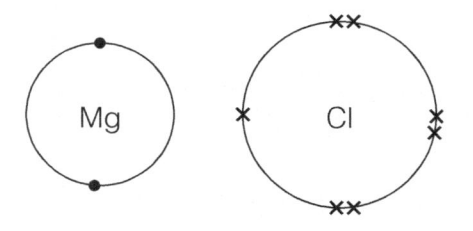

Magnesium forms an ionic compound with chlorine.

Describe what happens, in terms of the movement and transfer of electrons, when one atom of magnesium reacts with two atoms of chlorine. Give the formula of the compound. **[5 marks]**

_____

_____

_____

_____

_____

_____

0 5 A molecule of water is shown in the dot and cross diagram in **Figure 2**.

Figure 2

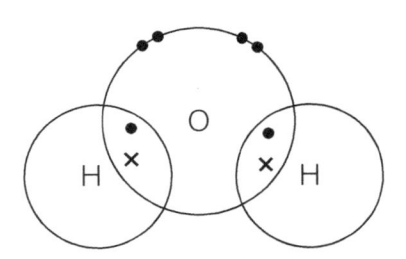

0 5 . 1 Suggest **one** advantage and **one** disadvantage of using the dot and cross diagram to show the structure of water. **[2 marks]**

Advantage: _____

Disadvantage: _____

Magnesium is a Group 2 metal. The structure of magnesium can be shown using the metallic model in **Figure 3.**

Figure 3

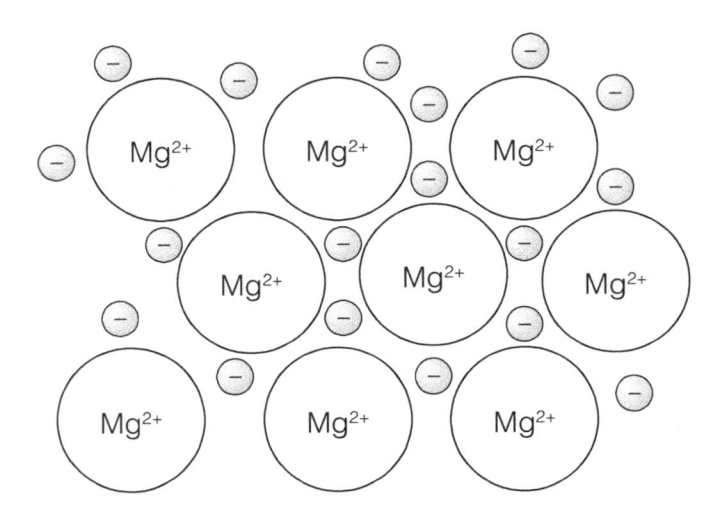

0 5 . 2 Suggest **one** advantage and **one** disadvantage of using the metallic model to show the structure of magnesium. **[2 marks]**

Advantage: _____

Disadvantage: _____

The structure of diamond can be shown using the ball and stick model in **Figure 4.**

Figure 4

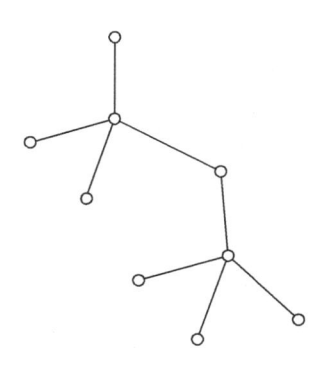

0 5 . 3 Suggest **one** advantage and **one** disadvantage of using the ball and stick model to show the structure of diamond.
[2 marks]

Advantage: _____

Disadvantage: _____

0 6 Carbon forms covalent bonds with four hydrogen atoms to form methane, CH_4.

0 6 . 1 Complete the dot and cross diagram in **Figure 5** to show the covalent bonds in one molecule of methane.

Only show the electrons in the outer shell of each atom.
[1 mark]

Figure 5

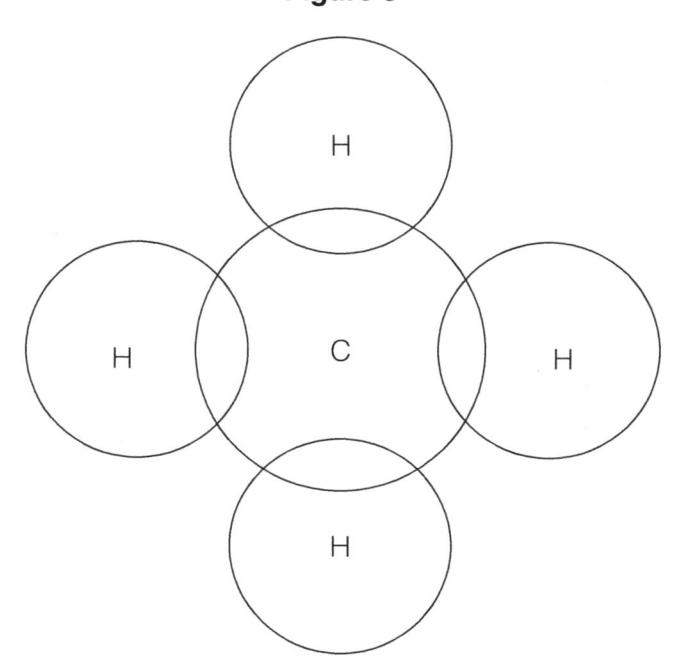

0 6 . 2 Methane is a gas at room temperature. Explain why methane has a low boiling point.
[2 marks]

_____

_____

_____

0 6 . 3 Methane does not conduct electricity when solid. Why does methane not conduct electricity?
[1 mark]

Tick **one** box.

The electrons are free to move. ☐

It has no charged particles that can move. ☐

The bonds between the molecules are weak. ☐

The bonds within the molecules are weak. ☐

0 7 Here are the relative atomic masses (A_r) of the elements in ammonium sulfate:

nitrogen = 14

hydrogen = 1

sulfur = 32

oxygen = 16

0 7 . 1 Calculate the relative formula mass (M_r) of ammonium sulfate, $(NH_4)_2SO_4$.
[2 marks]

_____

_____

_____

Relative formula mass = _____

0 7 . 2 What is the mass of two moles of ammonium sulfate?
[1 mark]

_____

_____

0 8 Fraser wanted to separate some pure water from a mixture of ink and water using the equipment shown in **Figure 6**.

Figure 6

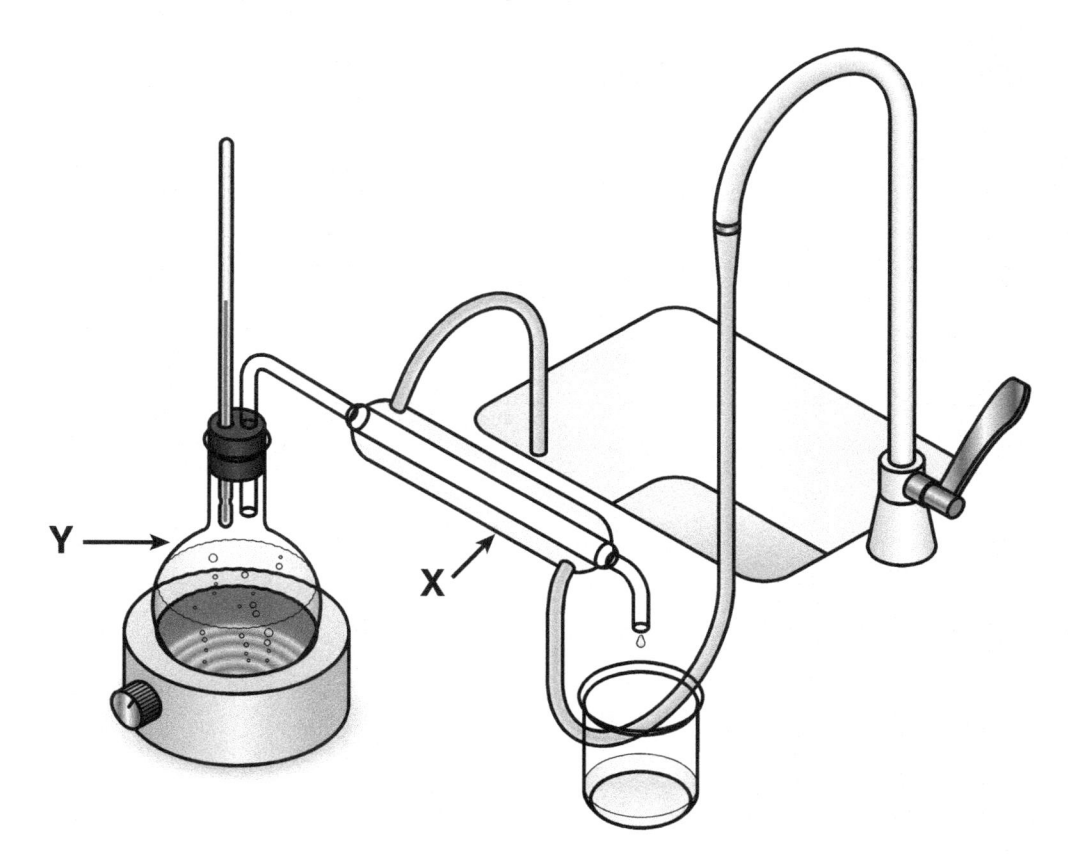

0 8 . 1 Name the pieces of apparatus labelled X and Y. [2 marks]

X: _____

Y: _____

0 8 . 2 Describe what happens at X and Y in terms of the changes of state. [2 marks]

At X:

_____

_____

At Y:

_____

_____

0 9 Ria is investigating how copper sulfate crystals can be formed.

Describe how a sample of pure, dry copper sulfate can be made from copper oxide. Include the name of the acid Ria should use. Describe what Ria should do. **[6 marks]**

1 0 Isaac is asked to produce 12.0 g of magnesium sulfate, $MgSO_4$.

The equation for the reaction is:

$MgO + H_2SO_4 \rightarrow MgSO_4 + H_2O$

Relative atomic masses (A_r):

magnesium = 24   oxygen = 16

sulfur = 32   hydrogen = 1

1 0 . 1 Calculate the mass of magnesium oxide that Isaac should react with excess dilute sulfuric acid to make 12.0 g of magnesium sulfate.

Give your answer to two significant figures. **[4 marks]**

The percentage yield for the reaction is 65%.

1 0 . 2 Calculate the mass of magnesium sulfate that Isaac actually produced. **[2 marks]**

_____

_____

1 0 . 3 Suggest **one** reason why the percentage yield for the reaction is less than 100%. **[1 mark]**

_____

_____

The percentage atom economy for a reaction is calculated using the equation:

$$\text{Percentage atom economy} = \frac{\text{Relative formula mass of desired product}}{\text{Sum of the relative formula mass of all the reactants}}$$

1 0 . 4 Calculate the percentage atom economy for this reaction.

Give your answer to three significant figures. **[3 marks]**

_____

_____

_____

_____

1 0 . 5 Why do scientists favour reactions with a high atom economy? **[1 mark]**

_____

_____

1 1 This question is about the electrolysis of copper sulfate solution.

Connie set up the equipment as shown in **Figure 7**.

Figure 7

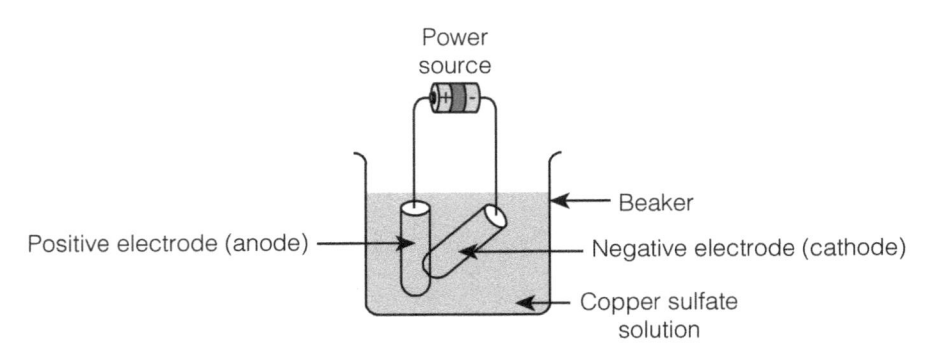

1 1 . 1 Identify the error in the way Connie set up the apparatus.

What would happen if Connie used the apparatus as shown in **Figure 7**? **[2 marks]**

_____

_____

_____

Connie changed the apparatus so the equipment was set up correctly.

The electrodes are copper.

The ionic equation for the reaction at the anode is:

$Cu \rightarrow Cu^{2+} + 2e^-$

1 1 . 2 Copper is oxidised at the anode. Explain why copper is oxidised in terms of electron transfer. **[1 mark]**

_____

The mass of the electrodes changes during the experiment.

1 1 . 3 Balance the ionic equation for the reaction that takes place at the cathode.

Explain what the ionic equation tells us about what happens in this reaction. **[3 marks]**

$Cu^{2+} + $ _____ $e^- \rightarrow Cu$

_____

_____

_____

_____

Connie repeated the experiment using different currents.

The mass of the anode was measured at the start. Each experiment was carried out for five minutes. The anode was then dried and the mass measured again. The results were recorded in **Table 3**.

Table 3

Current (A)	0.1	0.2	0.3	0.4	0.5
Mass of anode at start in g	2.34	2.38	2.35	2.31	2.32
Mass of anode at end in g	2.28	2.36	2.23	2.14	2.09
Change in mass in g	0.06	0.02	0.12		0.23

1 1 . 4 Calculate the change in mass when a current of 0.4 A was used. **[1 mark]**

The result for 0.2 A was anomalous.

1 1 . 5 Suggest what might have caused the anomalous result. **[1 mark]**

1 1 . 6 Suggest an improvement that Connie could make to the experiment to produce more accurate results. Give a reason for your answer. **[2 marks]**

1 1 . 7 Look at **Table 3**. Use the results to complete **Graph 1**.

Label the x axis.

Plot the results.

Draw a line of best fit. **[3 marks]**

Graph 1

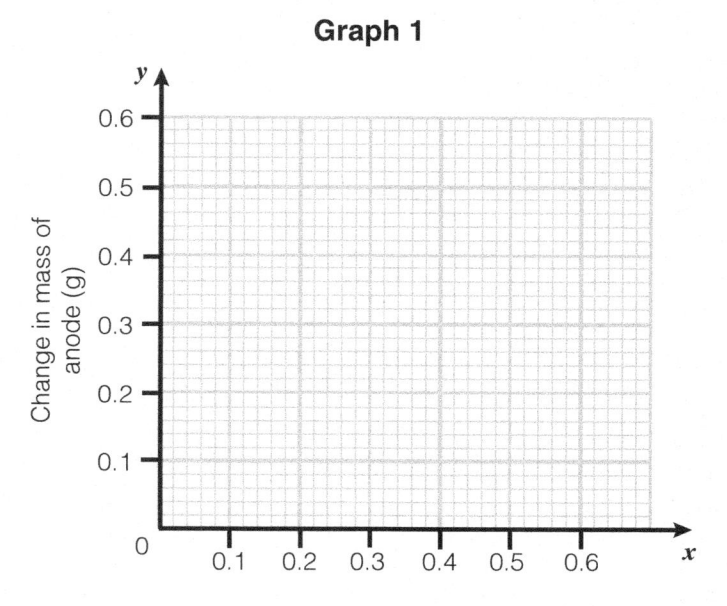

1 2 The Group 1 element potassium and the transition element copper are both metals.

Some of the properties of potassium and copper are shown in **Table 4**.

Table 4

	Potassium	Copper
Group	1	transition element
Melting point in °C	64	1085
Formula of chlorides	KCl	CuCl CuCl$_2$

Use your own knowledge and understanding and the data shown in **Table 4** to compare the properties of Group 1 metals and transition elements. **[6 marks]**

1 3 Hydrochloric acid is neutralised by potassium hydroxide to produce a salt and water.

1 3 . 1 Complete the equation for this reaction. **[1 mark]**

hydrochloric acid + _____ → _____ + water

Hydrochloric acid is a strong acid.

1 3 . 2 Explain the term 'strong acid'. **[2 marks]**

1 3 . 3 Give the ionic equation for the reaction between hydrochloric acid and potassium hydroxide. **[2 marks]**

Jacob carried out some titrations to find out the volume of sodium hydroxide that would be needed to neutralise 25.0 cm³ of hydrochloric acid.

1 3 . 4 Describe how Jacob should carry out the titration. Include the names of key pieces of equipment.

[6 marks]

___

___

___

___

___

___

___

Jacob carried out three titrations and recorded his results in **Table 5.**

Table 5

	1	2	3
Volume of potassium hydroxide in cm³	22.60	22.45	22.35

Concordant results are within 0.10 cm³ of each other.

1 3 . 5 Circle the concordant results shown in **Table 5**. **[1 mark]**

1 3 . 6 Use the concordant results to work out the mean volume of potassium hydroxide required to react with 25.0 cm³ of the hydrochloric acid solution. **[1 mark]**

___

___

This is the equation for the reaction between potassium hydroxide and hydrochloric acid:

KOH + HCl → KCl + H₂O

In another experiment, 22.70 cm³ of potassium hydroxide solution was required to neutralise 25.0 cm³ of hydrochloric acid.

The acid has a concentration of 0.20 mol / dm³.

13.7 Calculate the concentration of the potassium hydroxide solution.

Give your answer to three significant figures. **[4 marks]**

___

Becca has 100 cm³ of potassium hydroxide solution.

The concentration of the solution is 0.15 mol / dm³.

The relative formula mass (M_r) of potassium hydroxide = 56.

13.8 Calculate the mass of potassium hydroxide in 100 cm³ of this solution. **[2 marks]**

___

14 Figure 8 shows the displayed formula for the reaction between hydrogen and oxygen.

Figure 8

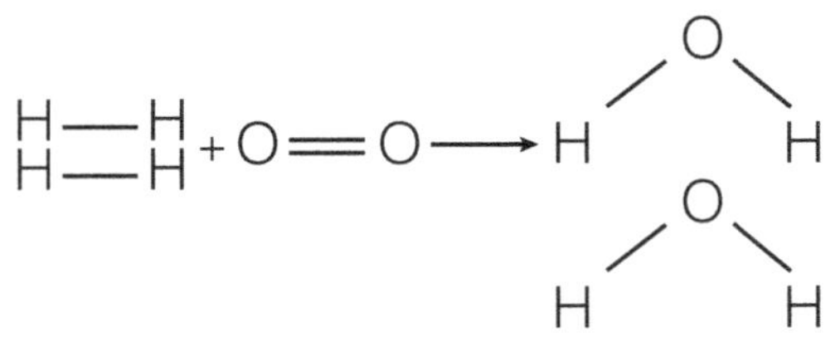

The atoms in a hydrogen molecule are held together by covalent bonds.

1 4 . 1 Describe what a covalent bond is. **[1 mark]**

1 4 . 2 Explain how a covalent bond holds the two hydrogen atoms together. **[2 marks]**

1 4 . 3 This question is about the reaction between hydrogen and oxygen to make water:

$2H_2 + O_2 \rightarrow 2H_2O$

The reaction is exothermic.

Complete **Figure 9** to show the reaction profile of this reaction.

Label:

- the reactants
- the products
- the energy given out (ΔH)
- the activation energy. **[4 marks]**

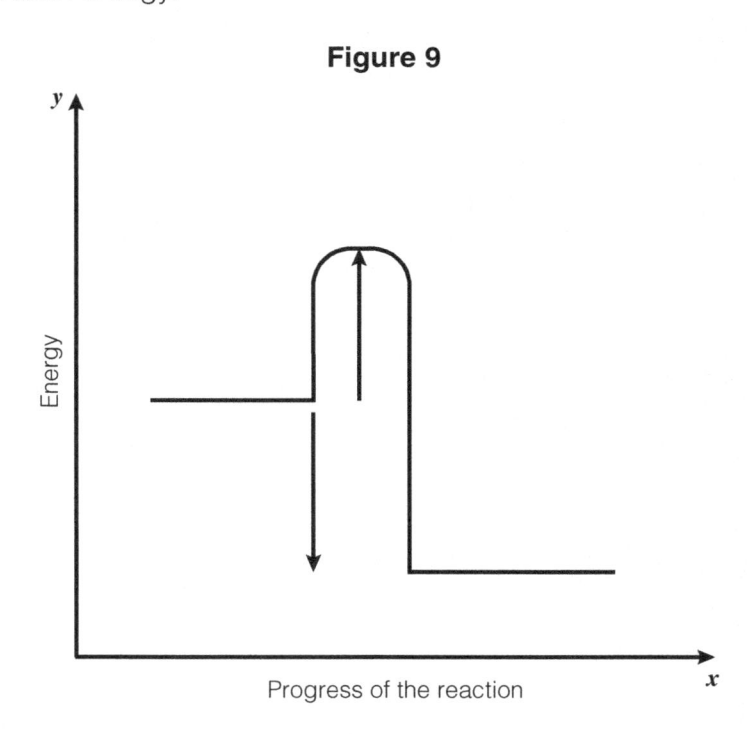

Figure 9

Table 6 shows the bond energies for a selection of covalent bonds.

Table 6

Bond	Energy in kJ / mol
H–H	436
O–H	463
O=O	495

1 4 . 4 Use **Table 6** and **Figure 8** to calculate the energy change for the reaction:

$2H_2 + O_2 \rightarrow 2H_2O$

[3 marks]

END OF QUESTIONS

GCSE (9–1)
Chemistry
Paper 2H

H

Higher tier
Time: 1 hour 45 minutes

Materials

For this paper you must have:
- a ruler
- a calculator
- the periodic table.

Instructions

- Answer **all** questions in the spaces provided.
- Do all rough work in this book. Cross through any work you do not want to be marked.

Information

- There are 100 marks available on this paper.
- The marks for questions are shown in brackets.
- You are expected to use a calculator where appropriate.
- You are reminded of the need for good English and clear presentation in your answers.
- When answering questions 07.2, 13.3, 14.6 and 15.1 you need to make sure that your answer:
  - is clear, logical, sensibly structured
  - fully meets the requirements of the question
  - shows that each separate point or step supports the overall answer.

Advice

In all calculations, show clearly how you work out your answer.

Name: _____

01 Anastasia carries out a flame test to identify the metal ions present in a sample of an unknown compound.

01.1 Describe how Anastasia should carry out the flame test. **[3 marks]**

01.2 The flame produced is lilac. Name the metal ion present. **[1 mark]**

01.3 Explain why it is important that the sample is not contaminated by sodium compounds. **[2 marks]**

In order to identify chemicals, chemists can use traditional methods of analysis like flame tests or modern methods of analysis such as flame emission spectroscopy.

01.4 Suggest **one** advantage and **one** disadvantage of using modern methods rather than traditional methods of analysis. **[2 marks]**

Advantage:

Disadvantage:

02 The products released into the atmosphere when fossil fuels are burnt can cause problems.

02.1 Draw **one** line from each pollutant to the environmental problem it causes. **[2 marks]**

Pollutant   **Environmental problem**

- Global dimming
- Carbon particles
- Global warming
- Nitrogen oxides
- Acid rain
- Thinning of the ozone layer

Petrol contains octane: C_8H_{18}. Octane is a hydrocarbon.

02.2 Explain why octane is described as a hydrocarbon. **[2 marks]**

02.3 Complete the equation below to show the complete combustion of octane. **[2 marks]**

C_8H_{18} + _____ O_2 → _____ CO_2 + $9H_2O$

02.4 If there is a limited supply of oxygen then incomplete combustion can occur. Carbon dioxide and water are produced. Name **one** other product. **[1 mark]**

03 This question is about cracking.

03.1 Hydrocarbon molecules can be cracked. What is an advantage of cracking hydrocarbon molecules? **[1 mark]**

Tick **one** box.

- It produces smaller, more useful alkanes and reactive alkenes. ☐
- It produces reactive alkanes and smaller, more useful alkenes. ☐
- It produces larger, more useful alkanes and reactive alkenes. ☐
- It produces smaller, more useful alkanes and unreactive alkenes. ☐

03.2 A hydrocarbon with 20 carbon atoms is cracked. Complete the equation to show what happens. **[1 mark]**

$C_{20}H_{42}$ → $C_{16}H_{34}$ + 2 _____

0 3 . 3 Give the general formula for an alkane molecule. [1 mark]

_____

0 4 This question is about ethanoic acid.

Ethanoic acid is a member of the carboxylic acid homologous series.

0 4 . 1 What is meant by the term 'homologous series'? [1 mark]

_____

0 4 . 2 Which of the structures in **Figure 1** shows the displayed formula of ethanoic acid? [1 mark]

Figure 1

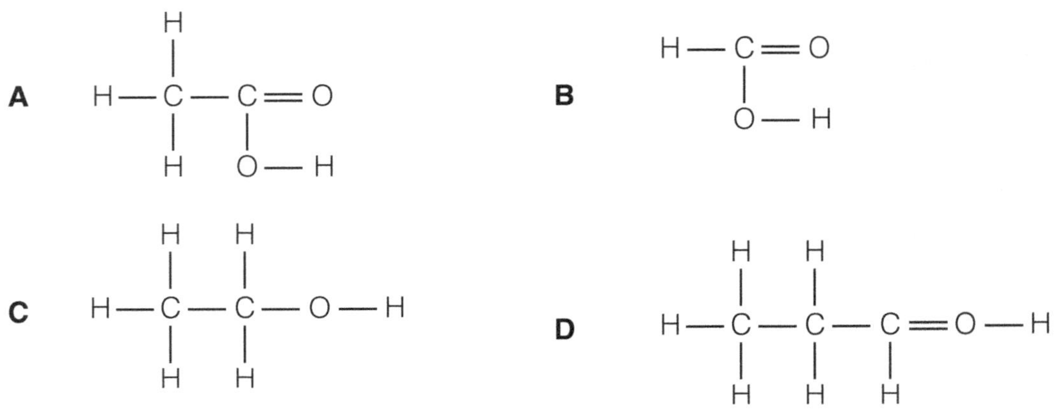

A
$$H-\underset{\underset{H}{|}}{\overset{\overset{H}{|}}{C}}-\underset{\underset{O-H}{|}}{C}=O$$

B
$$H-C=O$$
$$|$$
$$O-H$$

C
$$H-\underset{\underset{H}{|}}{\overset{\overset{H}{|}}{C}}-\underset{\underset{H}{|}}{\overset{\overset{H}{|}}{C}}-O-H$$

D
$$H-\underset{\underset{H}{|}}{\overset{\overset{H}{|}}{C}}-\underset{\underset{H}{|}}{\overset{\overset{H}{|}}{C}}-\underset{\underset{H}{|}}{C}=O-H$$

Tick **one** box.

A ☐   C ☐

B ☐   D ☐

0 4 . 3 Name the organic compound that can be oxidised to produce ethanoic acid. [1 mark]

Tick **one** box.

Ethene ☐

Ethane ☐

Ethanol ☐

Sodium ethanoate ☐

0 4 . 4 Name the ester produced when ethanol reacts with ethanoic acid. [1 mark]

_____

0 5 This question is about water.

0 5 . 1 What does the term 'potable' mean? [1 mark]

_____

0 5 . 2 Why is water passed through filter beds as part of the purification process? [1 mark]

_____

Chlorine is added to drinking water.

0 5 . 3 Give a reason why chlorine is added. [1 mark]

_____

Pure water can be produced from sea water by reverse osmosis or by distillation.

0 5 . 4 Explain what happens during the distillation of sea water. [2 marks]

_____

_____

_____

0 5 . 5 Give **one** disadvantage of using reverse osmosis to purify water. [1 mark]

_____

0 6 This question is about identifying metal hydroxides.

Barney added a solution of sodium hydroxide to a solution of copper (II) sulfate.

0 6 . 1 Give the name of the insoluble compound that was formed. [1 mark]

_____

0 6 . 2 Identify the colour of the precipitate made when a solution of copper (II) sulfate reacts with a solution of sodium hydroxide. [1 mark]

Tick **one** box.

Green	☐	Blue	☐
Brown	☐	Yellow	☐

Set A: Paper 2 25

0 6 . 3 Identify the ionic equation for the reaction between a solution of sodium hydroxide and a solution of copper (II) sulfate. **[1 mark]**

Tick **one** box.

$Cu^{2+} + OH^- \rightarrow CuOH$ ☐

$Cu^{2+} + 2OH^- \rightarrow Cu(OH)_2$ ☐

$Cu^{2+} + OH^- \rightarrow Cu(OH)_2$ ☐

$Cu^{2+} + 2OH^- \rightarrow CuOH$ ☐

0 7 This question is about identifying sodium halide compounds.

Parminder is given two white powders. One is sodium chloride and the other is sodium bromide. She wants to identify each of the compounds.

0 7 . 1 David suggests that Parminder could use a flame test to identify each of the compounds. Is he correct? Explain your answer. **[1 mark]**

_____

_____

0 7 . 2 Describe how Parminder could identify each of the compounds using silver nitrate solution.

Explain how the results could be used to identify each compound. **[5 marks]**

_____

_____

_____

_____

_____

_____

0 8 This question is about rusting.

0 8 . 1 A lot of effort is made to stop iron objects from rusting. Give **one** reason why rusting reactions are undesirable. **[1 mark]**

_____

0 8 . 2 Identify the **two** substances that must be present for an iron object to rust. **[2 marks]**

_____

_____

0 9 This question is about the manufacture of cars.

A manufacturing company produces a LCA for the cars it produces.

0 9 . 1 What does LCA stand for? **[1 mark]**

_____

0 9 . 2 Why is a LCA useful to customers? **[1 mark]**

_____

0 9 . 3 Other than the LCA, give **one** factor that customers might consider when buying a car. **[1 mark]**

_____

0 9 . 4 Which **one** of the steps below is **not** part of the LCA of a car? **[1 mark]**

Tick **one** box.

The top speed of the car ☐

How much of the car can be recycled ☐

The effect on the environment of extracting the raw materials to produce the car ☐

The amount of carbon dioxide produced in the manufacture of the car ☐

1 0 Shopping bags can be made from plastic or paper.

Explain the environmental impact of making bags from paper and from plastic. **[4 marks]**

_____

_____

_____

_____

_____

Set A: Paper 2

1 1 This question is about polymers. Polypropene is used to make ropes and crates.

1 1 . 1 Give the name of the monomer used to make polypropene. **[1 mark]**

_____

1 1 . 2 Polypropene is produced in an addition polymerisation reaction. Explain what happens in this reaction. **[2 marks]**

_____

_____

1 1 . 3 Poly(ethene) is another useful polymer. Complete the displayed formula for the production of poly(ethene) shown in **Figure 2**. **[2 marks]**

Figure 2

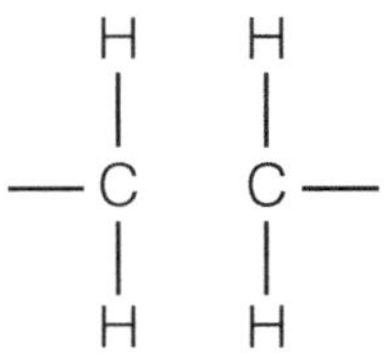

```
      H   H
      |   |
   —C   C—
      |   |
      H   H
```

1 2 This question is about fertilisers.

1 2 . 1 Ammonium sulfate is a useful fertiliser. Give the names of the **two** compounds that react together to form ammonium sulfate. **[2 marks]**

_____

_____

1 2 . 2 A fertiliser has the NPK value of 12:6:5.

What does this information show? **[1 mark]**

Tick **one** box.

The relative amounts of nitrogen, potassium and krypton ☐

The relative amounts of nitrogen, phosphorus and potassium ☐

How expensive it will be ☐

Whether it is soluble or not ☐

1 2 . 3 Why are NPK values useful to gardeners and farmers? **[1 mark]**

_____

1 3 This question is about hydrocarbons.

Ethane and ethene are hydrocarbons.

1 3 . 1 Give the molecular formula of ethene. **[1 mark]**

_____

1 3 . 2 Give the name of the homologous series that ethene belongs to. **[1 mark]**

_____

1 3 . 3 Describe a chemical test that a student could carry out to identify whether an unknown sample is ethane or ethene.

Include the names of any reagents and how you would use the results to identify the unknown compounds. **[4 marks]**

_____

_____

_____

_____

1 4 In industry, ammonia is produced using the Haber process. The equation for the reaction is:

$N_2 + 3H_2 \rightleftharpoons 2NH_3$

It is an exothermic reaction.

1 4 . 1 What is an exothermic reaction? **[1 mark]**

_____

Figure 3 shows a flow diagram of the Haber process.

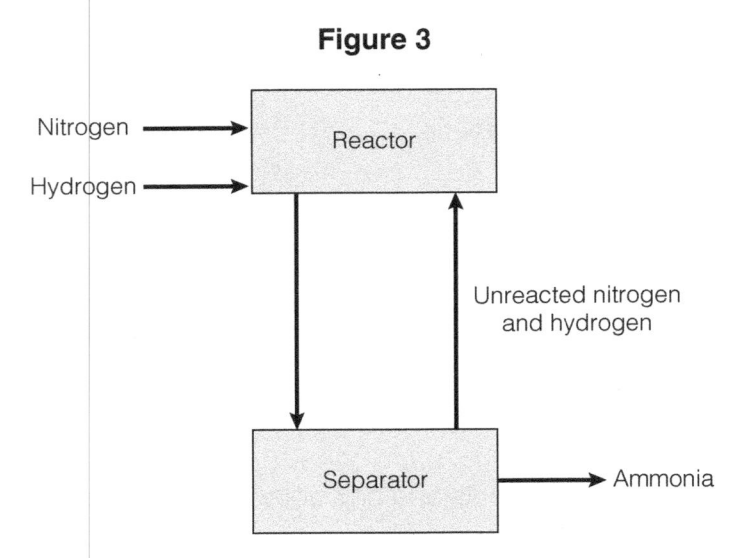

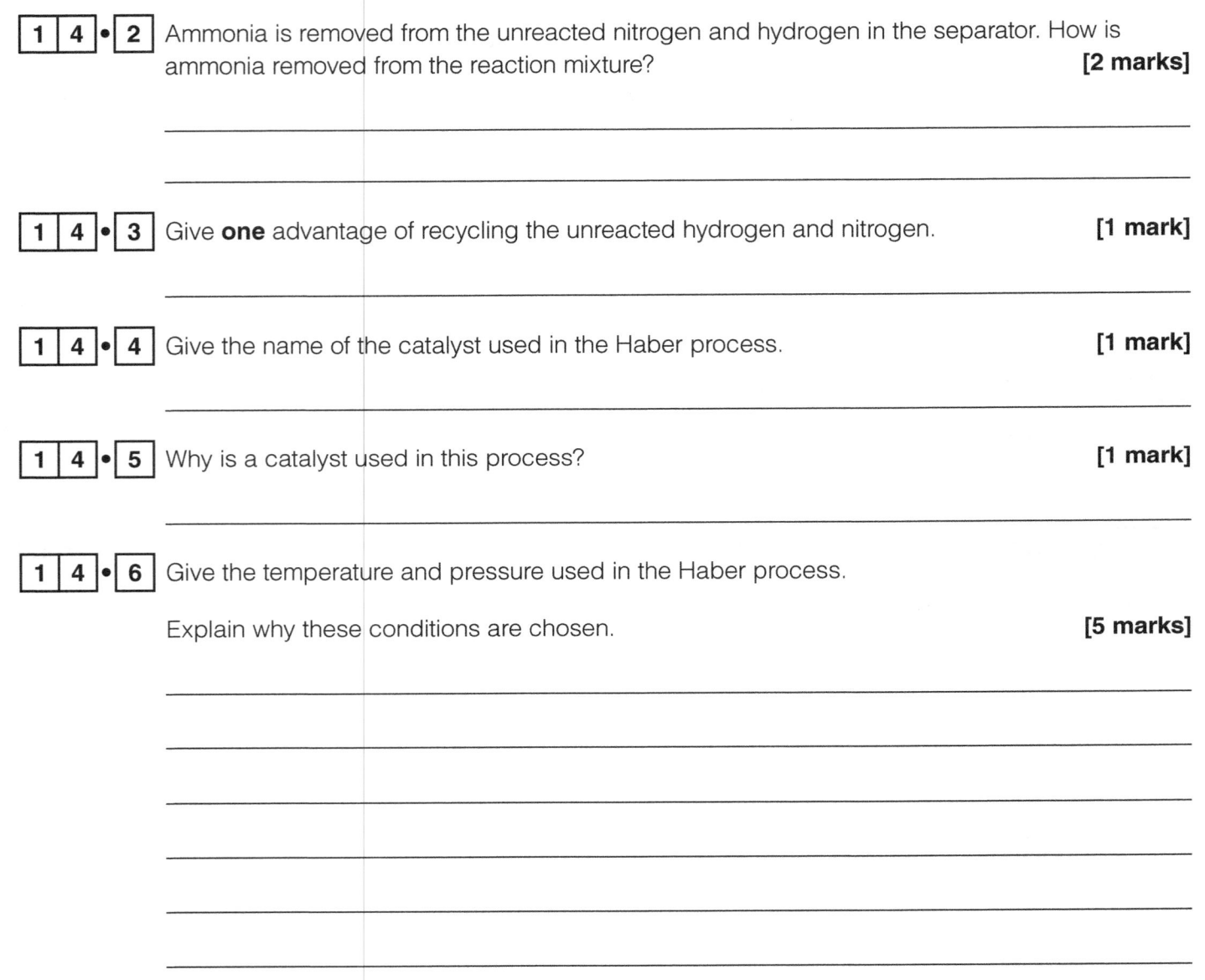

Figure 3

[Flow diagram: Nitrogen and Hydrogen enter the Reactor. Output goes to the Separator, which produces Ammonia and recycles Unreacted nitrogen and hydrogen back to the Reactor.]

1 4 . 2 Ammonia is removed from the unreacted nitrogen and hydrogen in the separator. How is ammonia removed from the reaction mixture? **[2 marks]**

1 4 . 3 Give **one** advantage of recycling the unreacted hydrogen and nitrogen. **[1 mark]**

1 4 . 4 Give the name of the catalyst used in the Haber process. **[1 mark]**

1 4 . 5 Why is a catalyst used in this process? **[1 mark]**

1 4 . 6 Give the temperature and pressure used in the Haber process.

Explain why these conditions are chosen. **[5 marks]**

15 This question is about chromatography.

15.1 Louise wants to produce a chromatogram of the ink in a felt-tip pen.

The ink in the pen is water-soluble. Describe how she could produce the chromatogram.
[5 marks]

15.2 Louise leaves the chromatogram in the beaker of water overnight.

Explain why this did not produce a useful chromatogram. **[1 mark]**

Another student carries out the experiment correctly and tests five felt-tip pens.

Figure 4 shows the chromatogram produced.

Figure 4

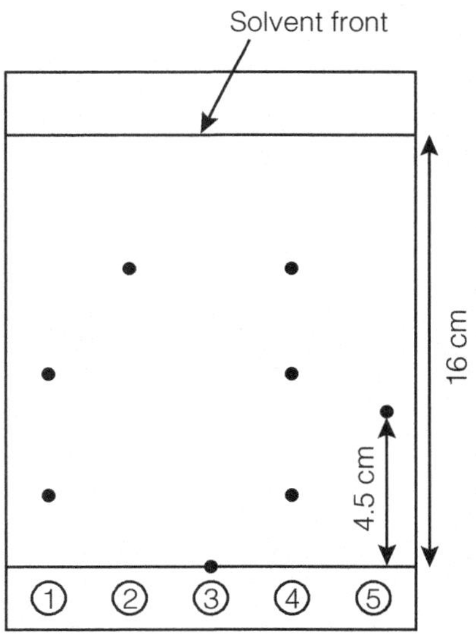

15.3 Calculate the R_f value for the ink in felt-tip pen 5. Give your answer to three significant figures.
[3 marks]

15.4 Look at **Figure 4**. Felt-tip pen 2 is a pure compound. How does the chromatogram show that it is a pure colour? **[1 mark]**

_____

15.5 The ink from felt-tip pen 3 does not move. What does this reveal about the ink in this pen? **[1 mark]**

_____

16 This question is about the rate of the reaction between magnesium and a solution of hydrochloric acid.

Mohammed set up the apparatus shown in **Figure 5**.

Figure 5

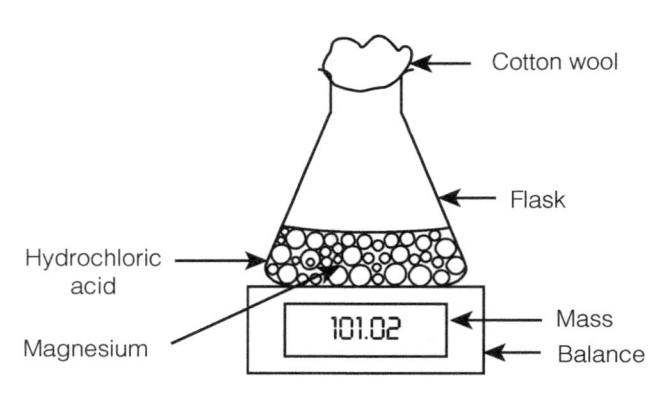

16.1 Complete and balance the equation for the reaction between magnesium and hydrochloric acid. **[2 marks]**

Mg + _____ → MgCl$_2$ + _____

16.2 Mohammed placed a piece of cotton wool in the top of the flask.

How does this make the results more accurate? **[1 mark]**

_____

_____

Table 1 shows the results of Mohammed's experiment.

Table 1

Time in s	Mass in g
0	200.0
20	188.0
40	176.0
60	168.0
80	161.0
100	155.0
120	151.0
140	148.0
160	145.0
180	142.0
200	140.0

1 6 . 3 Look at **Table 1**.

Use these results to complete **Graph 1**.

[4 marks]

Graph 1

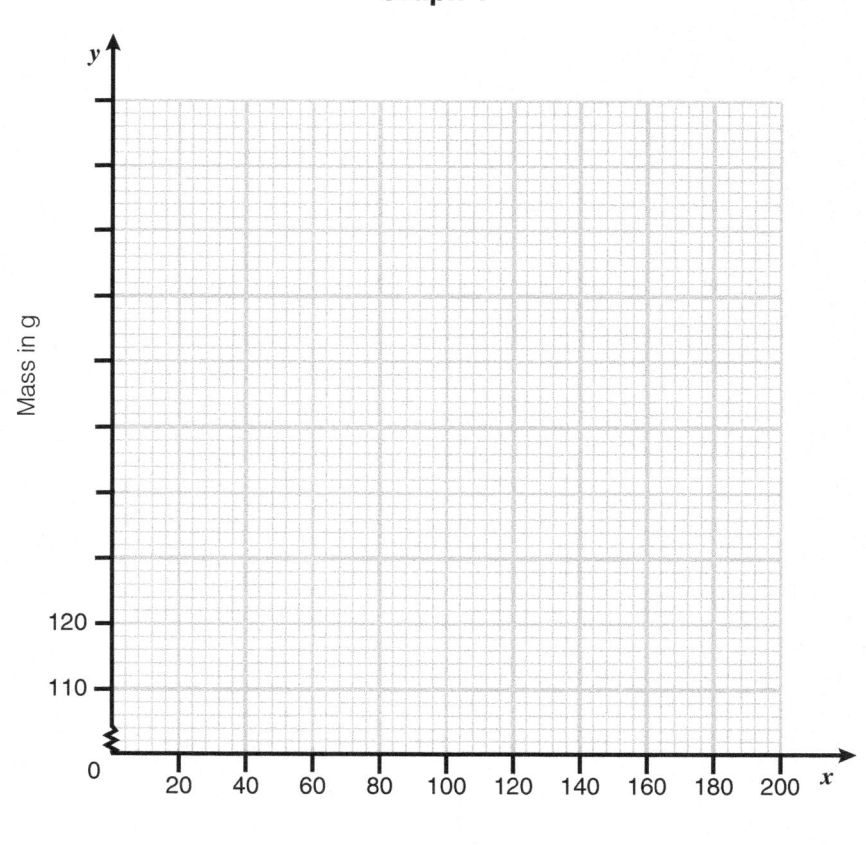

....

Byron carries out a similar experiment but this time measures the mass of hydrogen made.

The results of the experiment are shown in **Graph 2**.

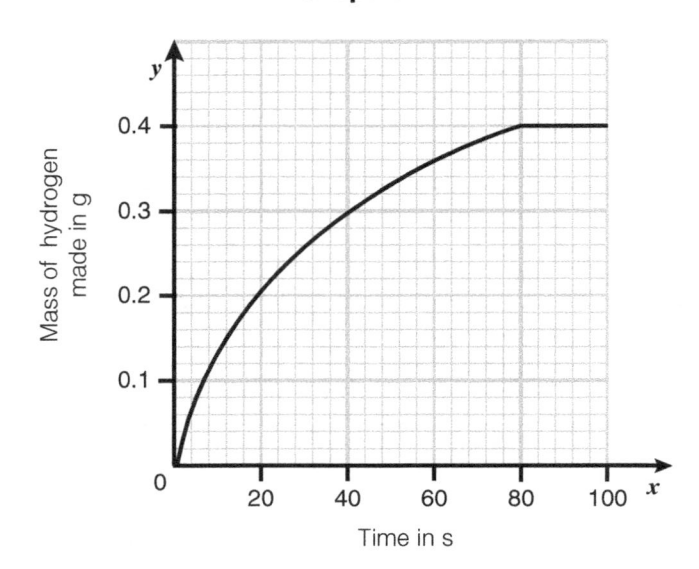

Graph 2

[Graph showing Mass of hydrogen made in g (y-axis, 0 to 0.4) against Time in s (x-axis, 0 to 100). Curve rises steeply then levels off at 0.4 g around 80 s.]

1 6 . 4 Describe how you could use **Graph 2** to calculate the rate of the chemical reaction at 50 seconds. **[2 marks]**

_____
_____

1 6 . 5 The rate of the reaction changes during the course of the reaction.

Describe how **Graph 2** shows how the rate of reaction changes. **[2 marks]**

_____
_____
_____

1 6 . 6 Explain, in terms of the particles involved, why increasing the temperature increases the rate of the reaction. **[4 marks]**

_____
_____
_____
_____
_____

END OF QUESTIONS

GCSE (9–1)
Chemistry
Paper 1H

H

Higher tier

Time: 1 hour 45 minutes

Materials

For this paper you must have:
- a ruler
- a calculator
- the periodic table.

Instructions

- Answer **all** questions in the spaces provided.
- Do all rough work in this book. Cross through any work you do not want to be marked.

Information

- There are 100 marks available on this paper.
- The marks for questions are shown in brackets.
- You are expected to use a calculator where appropriate.
- You are reminded of the need for good English and clear presentation in your answers.
- When answering questions 04, 05 and 10 you need to make sure that your answer:
  - is clear, logical, sensibly structured
  - fully meets the requirements of the question
  - shows that each separate point or step supports the overall answer.

Advice

In all calculations, show clearly how you work out your answer.

Name: _____

0 1 This question is about transition elements and Group 1 metals.

The transition elements and the Group 1 metals have some different physical and chemical properties.

Draw lines to match each group to its properties. **[4 marks]**

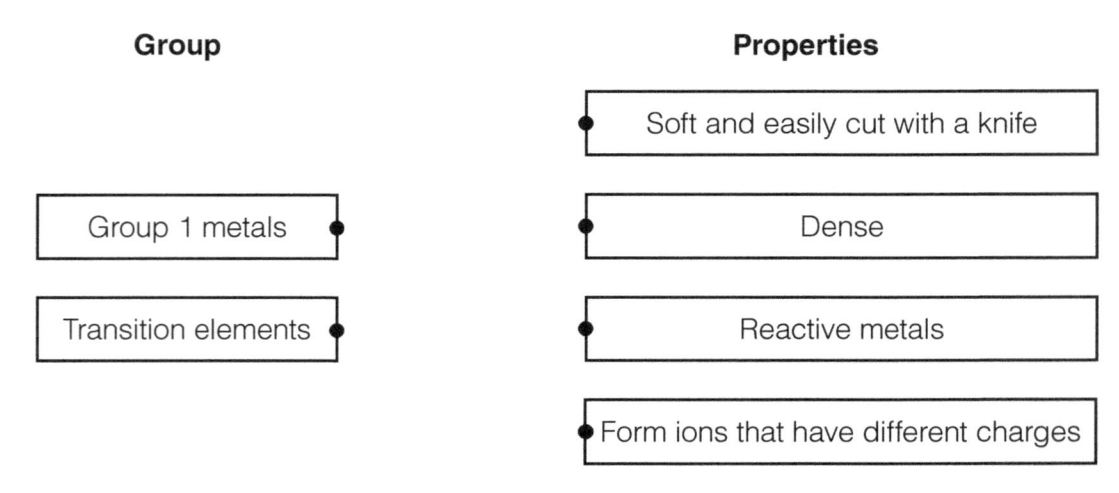

Group **Properties**

Soft and easily cut with a knife

Group 1 metals

Dense

Transition elements

Reactive metals

Form ions that have different charges

0 2 This question is about bromine.

Bromine has two common isotopes: bromine-79 and bromine-81.

0 2 . 1 Use the periodic table to identify the group that bromine belongs to.

[1 mark]

_____

0 2 . 2 What is the atomic number of bromine? **[1 mark]**

_____

0 2 . 3 Define the term 'isotope'. **[1 mark]**

_____

_____

0 2 . 4 In terms of subatomic particles, describe the similarities and the difference between an atom of bromine-79 and bromine-81. **[3 marks]**

_____

_____

_____

_____

0 3 An ion of fluorine has the symbol $^{19}_{9}F^-$.

0 3 . 1 Give the number of electrons in this ion. [1 mark]

Number of electrons = _____

0 3 . 2 Explain how ions can be formed. [2 marks]

_____

_____

0 3 . 3 Which group of the periodic table does fluorine belong to? Explain your answer in terms of fluorine's electron structure. [2 marks]

_____

_____

0 4 **Figure 1** shows an atom of the Group 1 element sodium and an atom of the Group 6 element oxygen. Only the outer shell of electrons is shown.

Figure 1

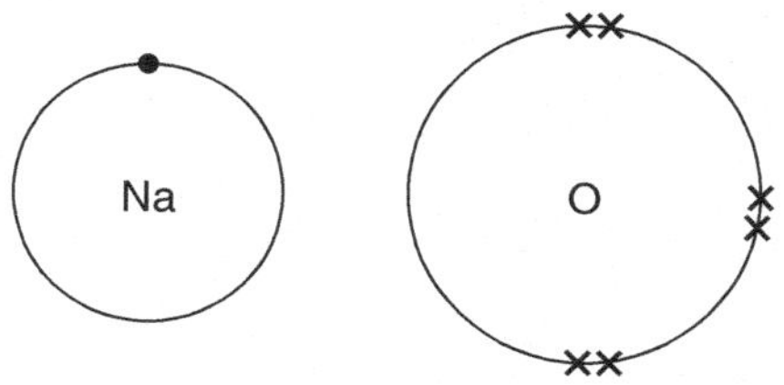

Sodium forms an ionic compound with oxygen.

Describe what happens when two atoms of sodium react with one atom of oxygen.

Give the formula of the compound formed and explain what happens in terms of electron transfer. [5 marks]

_____

_____

_____

_____

_____

0 5 The Group 1 element sodium conducts electricity when solid and when molten.

The Group 7 element chlorine does not conduct electricity when either solid or molten.

The compound sodium chloride does not conduct electricity when solid but does conduct electricity when molten.

Explain each of these statements. Include the types of bonding in each structure. **[6 marks]**

_____

_____

_____

_____

_____

_____

_____

0 6 Nitrogen molecules react with hydrogen molecules to form ammonia.

0 6 . 1 Complete and balance the equation to show the formation of ammonia. **[2 marks]**

N$_2$ + 3 _____ → _____ NH$_3$

A nitrogen atom forms covalent bonds with three hydrogen atoms to form ammonia, NH$_3$.

0 6 . 2 Complete the dot and cross diagram in **Figure 2** to show the covalent bonds in one molecule of ammonia.

Only show the electrons in the outer shell of each atom. **[2 marks]**

Figure 2

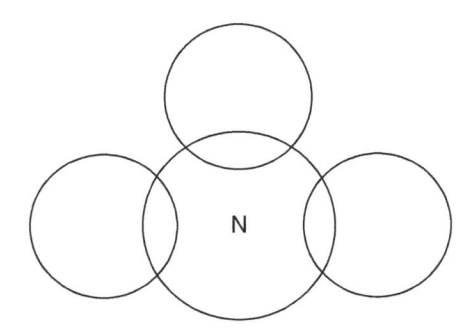

0 6 . 3 Explain why ammonia is a gas at room temperature. **[2 marks]**

_____

_____

0 7 This question is about the reactions of the halogens.

Table 1 shows the colour of the halogens in aqueous solution.

Table 1

Element	Colour of the aqueous solution
Fluorine	Colourless
Chlorine	Green
Bromine	Orange
Iodine	Brown

0 7 . 1 A redox reaction takes place when chlorine is added to a solution of potassium bromide.

Complete the word equation that sums up the reaction that takes place.

chlorine + potassium bromide → _____ + _____ **[1 mark]**

0 7 . 2 Look at **Table 1**. What is the colour of the final solution? **[1 mark]**

_____

0 7 . 3 Here is the ionic equation for the reaction between chlorine and potassium bromide:

$Cl_2 + 2Br^- \rightarrow Br_2 + 2Cl^-$

Which substance has been reduced in this reaction? Explain your answer. **[2 marks]**

_____

_____

07.4 Identify the products of the electrolysis of aqueous potassium chloride. **[1 mark]**

Tick **one** box.

Product at cathode	Product at anode	
hydrogen	oxygen	☐
hydrogen	chlorine	☐
potassium	chlorine	☐
potassium	oxygen	☐

08 A student is given the following information.

The formula of calcium hydroxide is Ca(OH)$_2$.

Relative atomic masses (A$_r$):

calcium = 40

hydrogen = 1

oxygen = 16

08.1 What is the relative formula mass of calcium hydroxide? **[2 marks]**

_____

_____

_____

Relative formula mass = _____

08.2 What is the mass of 2.5 moles of calcium hydroxide? **[1 mark]**

_____

_____

08.3 A student uses 50 cm³ of calcium hydroxide solution.

The concentration of the solution is 0.2 mol / dm³.

Calculate the mass of calcium hydroxide in 50 cm³ of this solution. **[2 marks]**

_____

_____

_____

0 9 This question is about copper.

Figure 3 shows the structure of copper.

Figure 3

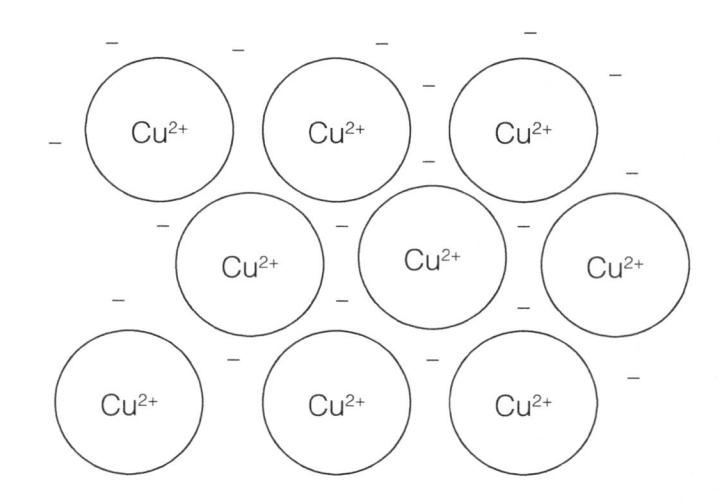

0 9 . 1 Name and describe the bonding in copper. [3 marks]

0 9 . 2 Explain why copper has a high melting point. [2 marks]

0 9 . 3 Explain why solid copper conducts electricity. [2 marks]

1 0 Atoms are made up of subatomic particles.

Name these subatomic particles and describe their relative mass and charge. **[6 marks]**

_____

_____

_____

_____

_____

1 1 Graphite is a form of the element carbon.

Figure 4 shows the structure of graphite.

Figure 4

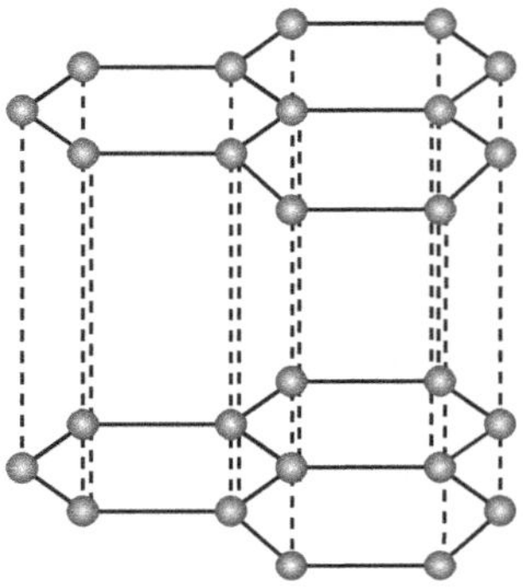

1 1 . 1 What do the circles and the solid lines in **Figure 4** represent? **[2 marks]**

Circles:
_____

Solid lines:
_____

1 1 . 2 Explain why graphite has a high melting point. **[2 marks]**

_____

_____

1 1 . 3 Explain why graphite conducts electricity. **[2 marks]**

_____

_____

1 2 The Avogadro constant has a value of 6.02×10^{23}.

Use the periodic table to answer the questions.

1 2 . 1 How many atoms are present in 24 g of Mg? [1 mark]

_____

_____

1 2 . 2 How many ions are present in 11.5 g of Na⁺? [1 mark]

_____

_____

1 3 This question is about the states of matter: solid, liquid and gas.

The three states of matter can be represented by the particle model.

1 3 . 1 Give **one** limitation of the particle model. [1 mark]

_____

1 3 . 2 Describe how the particles are arranged and how the particles move in each of the three states of matter. [6 marks]

_____

_____

_____

_____

_____

_____

_____

1 4 Nitric acid is neutralised by sodium hydroxide to produce a salt and water.

1 4 . 1 Complete the equation for this reaction.

nitric acid + _____ → _____ + water **[1 mark]**

Nitric acid is a strong acid.

1 4 . 2 Give the formula of nitric acid. **[1 mark]**

_____

1 4 . 3 Give the ionic equation for the reaction between nitric acid and sodium hydroxide. **[2 marks]**

_____

In another experiment, 28.75 cm³ of potassium hydroxide solution was required to neutralise 25.0 cm³ of sulfuric acid.

The equation for the reaction between potassium hydroxide and sulfuric acid is:

$2KOH + H_2SO_4 \rightarrow K_2SO_4 + 2H_2O$

The acid has a concentration of 0.10 mol / dm³.

1 4 . 4 Calculate the concentration of the potassium hydroxide solution.

Give your answer to three significant figures. **[4 marks]**

_____
_____
_____
_____
_____

1 5 Figure 5 shows the displayed formula for the reaction between hydrogen and chlorine.

Figure 5

$$H—H + Cl—Cl \rightarrow \begin{matrix} H—Cl \\ H—Cl \end{matrix}$$

1 5 . 1 Identify the type of chemical bond between the hydrogen atoms. **[1 mark]**

_____

Table 2 shows the bond energies for some covalent bonds.

Table 2

Bond	Energy in kJ / mol
H—H	436
H—Cl	432

The energy change for the reaction $H_2 + Cl_2 \rightarrow 2HCl$ is found to be −185 kJ / mol.

1 5 . 2 Use **Table 2** and **Figure 5** to calculate the bond energy for the Cl—Cl bond. **[4 marks]**

1 6 This question is about the reaction that happens when calcium carbonate is heated.

The calcium carbonate decomposes to form calcium oxide and carbon dioxide:

$CaCO_3 \rightarrow CaO + CO_2$

The reaction is endothermic.

1 6 . 1 What does the term 'endothermic' mean? **[1 mark]**

___

1 6 . 2 Complete **Figure 6** to show the reaction profile of this reaction.

Label:

- the reactants
- the products
- the energy change (ΔH)
- the activation energy. **[4 marks]**

Figure 6

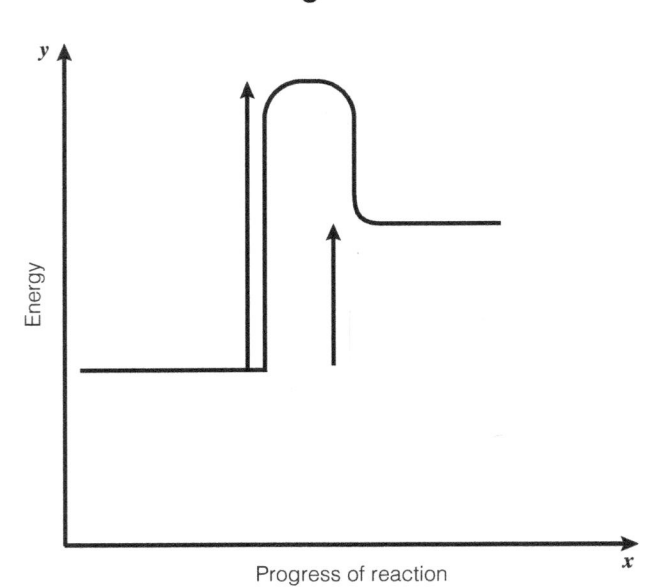

1 6 . 3 What does the term 'activation energy' mean? **[1 mark]**

___

1 7 Petra is asked to produce 7.4 g of magnesium nitrate, Mg(NO₃)₂.

The equation for the reaction is:

MgO + 2HNO₃ → Mg(NO₃)₂ + H₂O

Relative atomic masses (A$_r$):

magnesium = 24

nitrogen = 14

oxygen = 16

hydrogen = 1

1 7 . 1 Calculate the mass of magnesium oxide that Petra should react with excess dilute nitric acid to make 7.4 g of magnesium nitrate.

Give your answer to two significant figures. **[4 marks]**

The percentage yield for the reaction is 70%.

1 7 . 2 Calculate the mass of magnesium nitrate that Petra actually produced.

Give your answer to two significant figures. **[2 marks]**

17.3 Suggest why the actual yield is less than the theoretical yield. [1 mark]

The percentage atom economy of a reaction is calculated using the equation:

$$\text{Percentage atom economy} = \frac{\text{Relative formula mass of desired product}}{\text{Sum of the relative formula mass of all the reactants}}$$

17.4 Calculate the percentage atom economy for the production of magnesium nitrate.

Give your answer to three significant figures. [3 marks]

17.5 Why do scientists choose reactions with a high atom economy? [1 mark]

END OF QUESTIONS

GCSE (9–1)
Chemistry
Paper 2H

H

Higher tier
Time: 1 hour 45 minutes

Materials

For this paper you must have:
- a ruler
- a calculator
- the periodic table.

Instructions

- Answer **all** questions in the spaces provided.
- Do all rough work in this book. Cross through any work you do not want to be marked.

Information

- There are 100 marks available on this paper.
- The marks for questions are shown in brackets.
- You are expected to use a calculator where appropriate.
- You are reminded of the need for good English and clear presentation in your answers.
- When answering questions 6.4 and 10.2 you need to make sure that your answer:
  - is clear, logical, sensibly structured
  - fully meets the requirements of the question
  - shows that each separate point or step supports the overall answer.

Advice

In all calculations, show clearly how you work out your answer.

Name: _____

01 This question is about the problems caused by burning fossil fuels.

Identify the main problem caused by each of these pollutants. **[2 marks]**

Carbon dioxide: _____

Sulfur oxides: _____

02 This question is about crude oil and the cracking of hydrocarbons.

Alkanes are obtained from crude oil.

02.1 Describe how crude oil was formed. **[2 marks]**

_____

_____

02.2 Give the name of the process by which groups of the hydrocarbons in crude oil are separated. **[1 mark]**

_____

Many useful substances such as petrol can be extracted from crude oil.

Petrol contains nonane, C_9H_{20}. Nonane is a hydrocarbon.

02.3 Name the elements found in nonane. **[2 marks]**

_____

_____

Figure 1 shows the displayed formula of nonane.

Figure 1

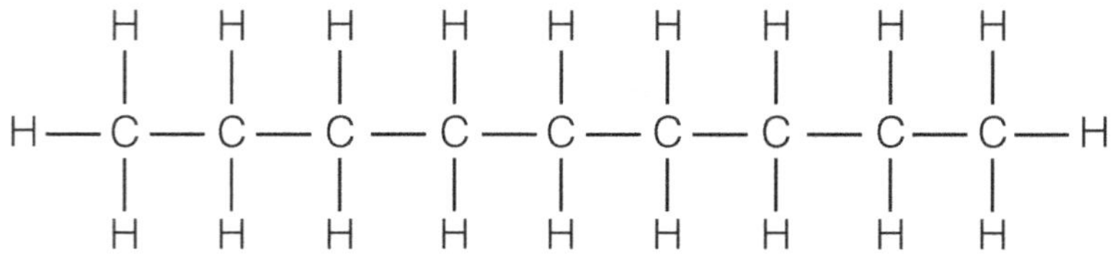

H—C(H)(H)—C(H)(H)—C(H)(H)—C(H)(H)—C(H)(H)—C(H)(H)—C(H)(H)—C(H)(H)—C(H)(H)—H

02.4 What do the lines in the displayed formula represent? **[1 mark]**

_____

02.5 Complete the equation to show the complete combustion of nonane. **[2 marks]**

C₉H₂₀ + _____ O₂ → _____ CO₂ + 10H₂O

Incomplete combustion of fuels can also occur.

02.6 Describe why incomplete combustion occurs. **[1 mark]**

_____

The cracking of hydrocarbons makes more useful products. One of the products of cracking is a shorter chain alkane.

02.7 Give the general formula for an alkane molecule. **[1 mark]**

_____

02.8 Explain why shorter alkane molecules are more useful than longer alkane molecules. **[2 marks]**

_____

_____

02.9 A hydrocarbon with 22 carbon atoms is cracked.

Complete the equation to show what happens. **[1 mark]**

C₂₂H₄₆ → C₁₉H₄₀ + _____

02.10 Name the alkene produced in this cracking reaction. **[1 mark]**

_____

02.11 Give **one** use of alkenes. **[1 mark]**

_____

03 This question is about identifying chemicals.

03.1 Penny conducted some flame tests.

Draw a line to link each metal ion to the flame colour it produces. **[2 marks]**

Metal ion	Flame colour
	Green
	Orange-red
Calcium	
	Crimson-red
Copper	
	Lilac
	Yellow

03.2 Precipitates are formed when metal ions react with hydroxide ions.

Identify the colour of the precipitate that each of these metal ions forms with hydroxide ions. **[2 marks]**

Metal ion   **Precipitate colour**

Al^{3+} _____

Fe^{3+} _____

03.3 Calcium (Ca^{2+}) ions react with hydroxide ions to form a white precipitate of calcium hydroxide, $Ca(OH)_2$.

Give the state of the calcium hydroxide formed in this reaction. **[1 mark]**

_____

03.4 Complete and balance the equation for this reaction. **[2 marks]**

_____ + _____ $OH^- \rightarrow$ _____

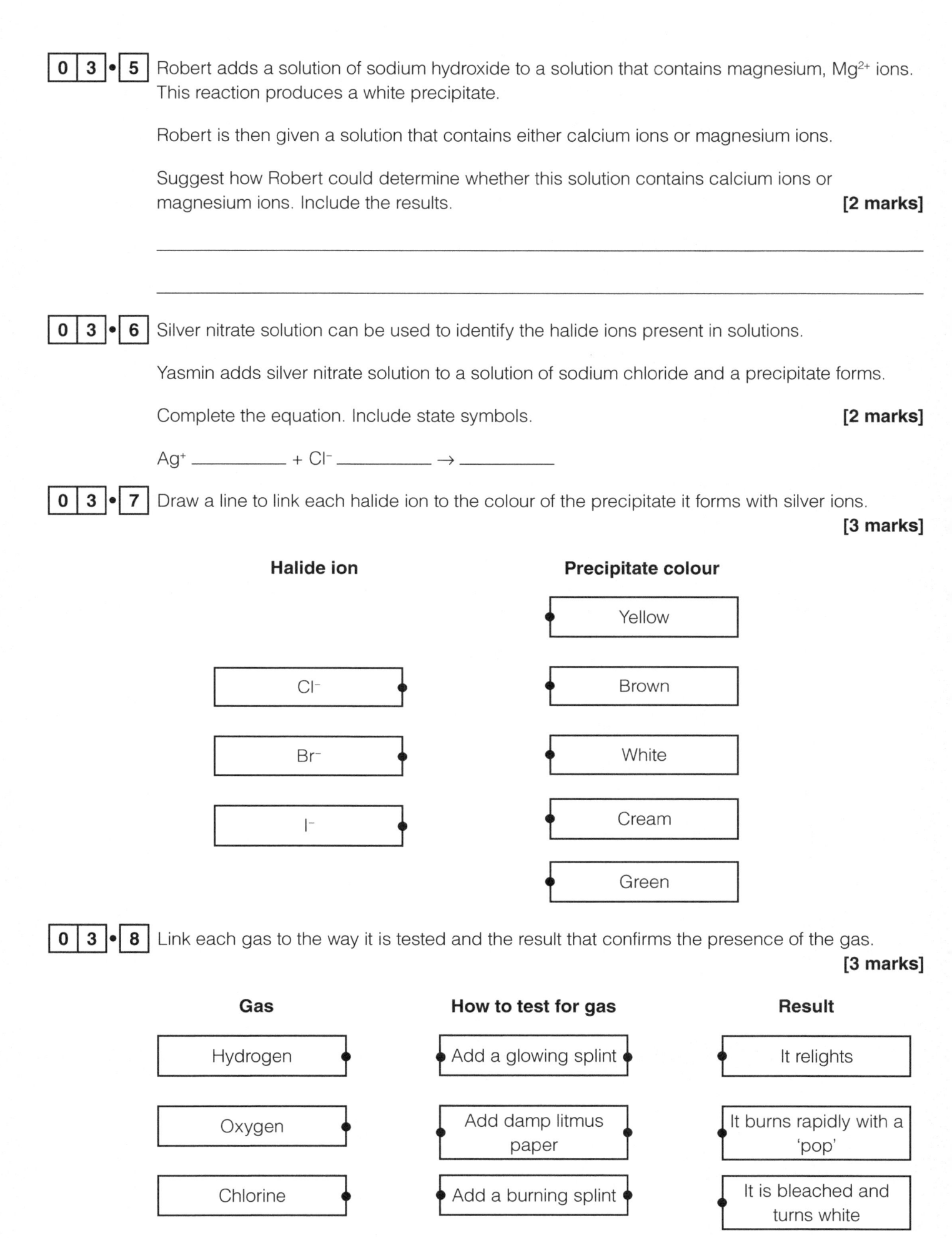

03.5 Robert adds a solution of sodium hydroxide to a solution that contains magnesium, Mg^{2+} ions. This reaction produces a white precipitate.

Robert is then given a solution that contains either calcium ions or magnesium ions.

Suggest how Robert could determine whether this solution contains calcium ions or magnesium ions. Include the results. **[2 marks]**

_____

_____

03.6 Silver nitrate solution can be used to identify the halide ions present in solutions.

Yasmin adds silver nitrate solution to a solution of sodium chloride and a precipitate forms.

Complete the equation. Include state symbols. **[2 marks]**

Ag⁺ _____ + Cl⁻ _____ → _____

03.7 Draw a line to link each halide ion to the colour of the precipitate it forms with silver ions. **[3 marks]**

Halide ion	Precipitate colour
	Yellow
Cl⁻	Brown
Br⁻	White
I⁻	Cream
	Green

03.8 Link each gas to the way it is tested and the result that confirms the presence of the gas. **[3 marks]**

Gas	How to test for gas	Result
Hydrogen	Add a glowing splint	It relights
Oxygen	Add damp litmus paper	It burns rapidly with a 'pop'
Chlorine	Add a burning splint	It is bleached and turns white

0 3 . 9 Describe how to test if an unknown solid contains carbonate ions.

Include the names of any reagents. **[3 marks]**

_____

_____

_____

0 4 This question is about the decomposition of hydrogen peroxide to form water and oxygen:

$$2H_2O_2 \rightarrow 2H_2O + O_2$$

The reaction is exothermic.

0 4 . 1 Complete **Figure 2** to show the reaction profile of this reaction.

Label:

- the energy given out (ΔH)
- the activation energy. **[2 marks]**

Figure 2

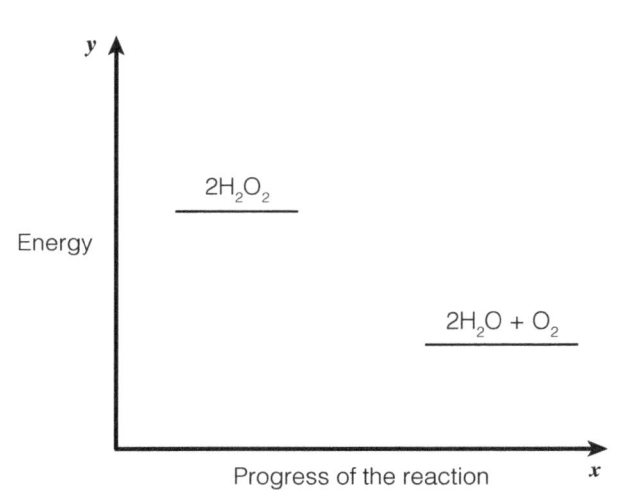

0 4 . 2 The reaction is repeated and a catalyst of manganese (IV) oxide is added.

Add a line to **Figure 2** to show the reaction pathway for the catalysed reaction. **[1 mark]**

0 4 . 3 Explain the role of the catalyst in this reaction. **[2 marks]**

_____

_____

0 4 . 4 Suggest why the catalyst is not included in the equation for the reaction. **[1 mark]**

_____

0 5 This question is about organic compounds.

Propanol is a member of the alcohol homologous series.

0 5 . 1 Give the functional group present in propanol. **[1 mark]**

_____

0 5 . 2 Look at **Figure 3**.

Figure 3

A H—C(H)(H)—C(H)(H)—C(H)(H)—C(H)(H)—O—H

B H—C(H)(H)—C(H)(H)—C(H)(H)—O—H

C H—C(H)(H)—C(H)(H)—C(=O)—OH

D H—C(H)(H)—C(H)(H)—C(H)(H)—H

Which of the structures in **Figure 3** shows the displayed formula of propanol? **[1 mark]**

Tick **one** box.

A ☐

B ☐

C ☐

D ☐

0 5 . 3 Complete and balance the equation of the complete combustion of propanol. **[2 marks]**

C₃H₇OH + _____ → _____ CO₂ + _____

Figure 4 shows the displayed formulae of four organic compounds.

Figure 4

A
```
        H
        |
    H — C — C = O
        |   |
        H   O — H
```

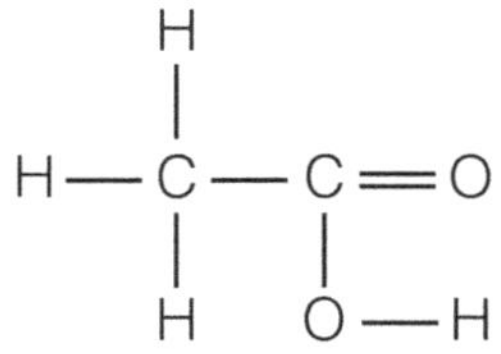

B
```
        H   H   H
        |   |   |
    H — C — C — C — H
        |   |   |
        H   H   H
```

C
```
    H       H
    |       |
    C = C — C — H
    |   |   |
    H   H   H
```

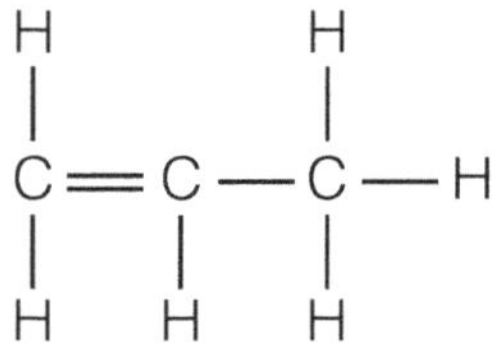

D
```
        H   O—H  H
        |   |   |
    H — C — C — C — H
        |   |   |
        H   H   H
```

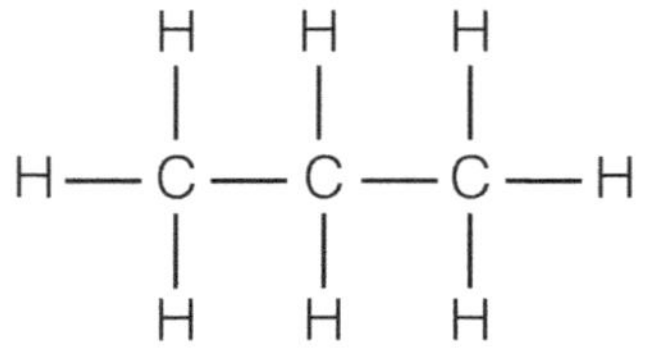

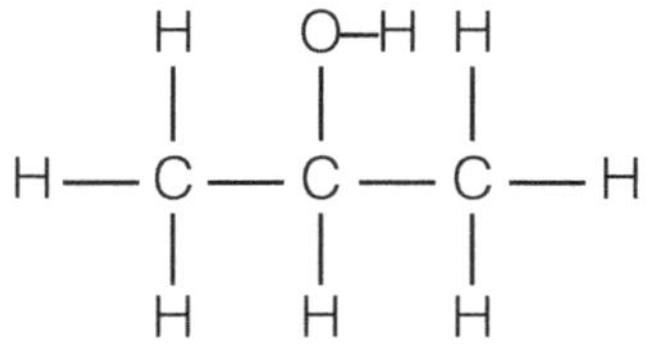

0 5 . 4 Name each of the compounds. **[4 marks]**

A _____

B _____

C _____

D _____

05.5 Look at **Figure 4**.

Match each of the compounds in **Figure 4** to the correct statement.

Insert A, B, C or D next to each statement. Each letter can be used once, more than once or not at all. **[4 marks]**

A solution of this compound turns universal indicator orange. ☐

This hydrocarbon decolourises bromine water. ☐

This compound forms a neutral solution with water and reacts with sodium to produce hydrogen. ☐

This compound has the general formula C_nH_{2n+2}. ☐

06 This question is about metals.

06.1 Draw a line to link each metal alloy to its property and use. **[3 marks]**

Metal alloy	Property	Use
Alloys of gold	Low density	Cutlery
Stainless steel	Hard and resists corrosion	Jewellery
Aluminium alloys	Unreactive	Aeroplanes

06.2 Explain what an alloy is and why most metals are alloyed for everyday use. **[2 marks]**

_____

_____

Water must be present for iron objects to rust.

06.3 Give the name of the gas, found in air, that must also be present for iron to rust. **[1 mark]**

_____

Set B: Paper 2

Jeremiah sets up an experiment to investigate the factors that affect how quickly an iron object rusts.

Figure 5 shows his experiment.

Figure 5

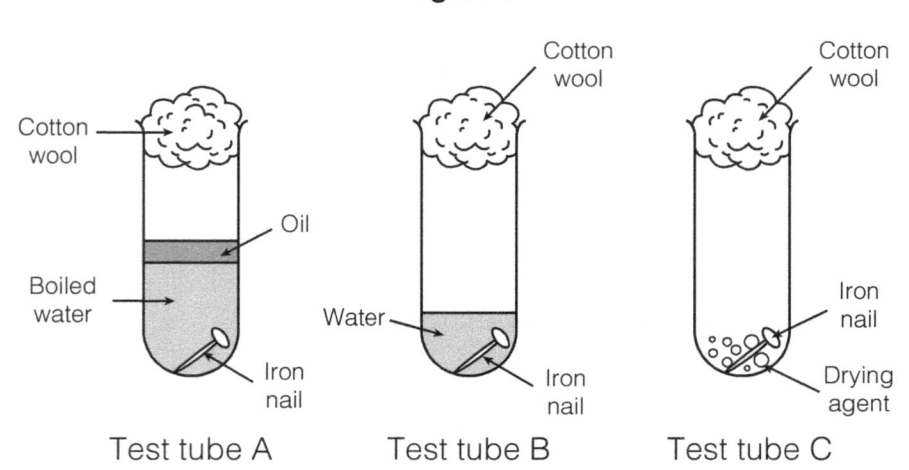

Jeremiah leaves the experiment for one week and then checks the iron nails for signs of rusting.

After one week:

- the nail in test tube A shows no change
- the nail in test tube B shows several brown spots
- the nail in test tube C shows no change.

0 6 . 4 Explain these observations. **[6 marks]**

Jeremiah reads that valuable iron objects can be protected from rusting by sacrificial protection.

0 6 . 5 Suggest a metal that could be used to protect the iron and explain how sacrificial protection works. **[3 marks]**

_____

_____

_____

0 7 This question is about the reaction between a solution of dilute hydrochloric acid and magnesium.

Explain, in terms of the particles involved, how increasing the concentration of the dilute hydrochloric acid affects the rate of the reaction. **[3 marks]**

_____

_____

_____

0 8 The carbon footprint of a product is the total amount of greenhouse gases emitted over the full life cycle of the product.

0 8 . 1 Give the name of a greenhouse gas. **[1 mark]**

_____

0 8 . 2 Which of these actions could reduce the carbon footprint of a product? **[2 marks]**

Tick **two** boxes.

Carbon capture and storage ☐

Increasing the price of the object ☐

Decreasing the atom economy of the reaction that makes the product ☐

Tree planting ☐

Using fossil fuels rather than alternative energy sources ☐

0 8 . 3 Give **one** problem that could prevent a reduction in the amount of greenhouse gases being produced. **[1 mark]**

0 9 This question is about metals.

0 9 . 1 What is the name of the rocks that contain metals in high enough quantities that it is economically worthwhile to extract the metal from the rock? **[1 mark]**

0 9 . 2 Suggest why scientists are trying to find methods of extracting copper from rocks that contain the metal in very low quantities. **[1 mark]**

0 9 . 3 Give a disadvantage of traditional methods of extracting copper compared with modern methods of extraction. **[1 mark]**

0 9 . 4 Describe how copper is extracted in bioleaching. **[1 mark]**

0 9 . 5 Describe how copper is obtained in phytomining. **[2 marks]**

0 9 . 6 Flame emission spectroscopy can be used to analyse solutions that contain metal ions.

A sample is placed in a flame. The light produced is passed through a spectroscope.

What useful information is obtained using this technique? **[2 marks]**

1 0 The atmosphere of the Earth has remained nearly constant over the past 200 million years.

1 0 . 1 Describe the present-day atmosphere of the Earth.

Include the names of the gases present and their percentage abundances. **[3 marks]**

_____
_____
_____
_____

1 0 . 2 Describe how scientists believe that the Earth's atmosphere has changed over time.

Include the names of the gases and details of how and why their concentrations have changed.
[6 marks]

_____
_____
_____
_____
_____
_____
_____
_____

1 1 This question is about types of plastic.

Figure 6 shows the structures of two polymers: melamine and poly(ethene).

Figure 6

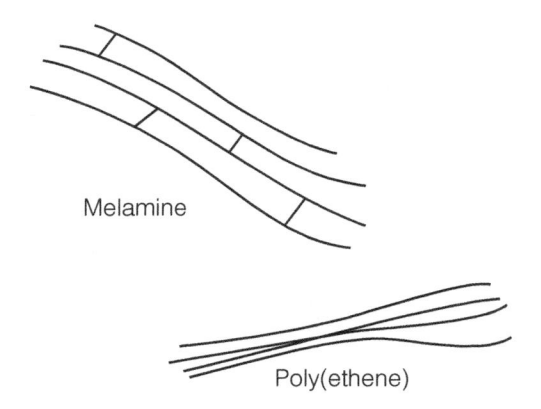

Melamine

Poly(ethene)

1 1 . 1 Explain how poly(ethene) is made.

Include the name of the type of reaction. **[3 marks]**

_____

_____

_____

1 1 . 2 Which plastic would be most suitable for producing a jug that could hold boiling water?

Explain your answer. **[3 marks]**

_____

_____

_____

END OF QUESTIONS

SET A: PAPER 1

01.1 The forces between the rubidium ions and the delocalised electrons are weaker than the forces between the potassium ions and the delocalised electrons **(1)**.

01.2 2Na(s) + Cl$_2$(g) → 2NaCl(s)
(1 mark for balancing and correct formulae; 1 mark for state symbols)

01.3 The ions cannot move **(1)**.

TOTAL MARKS FOR QUESTION 1 = 4

02.1 potassium bromide + iodine **(1)**
02.2 Brown **(1)**
02.3 2Br⁻ **(1)** + I$_2$ **(1)**
02.4 Hydrogen and bromine **(1)**

TOTAL MARKS FOR QUESTION 2 = 5

03.1 Number of protons = 15 **(1)**
Number of neutrons = 16 **(1)**
Number of electrons = 15 **(1)**
03.2 It has five electrons in its outer shell **(1)**.

TOTAL MARKS FOR QUESTION 3 = 4

04 **This is a model answer, which would gain all 5 marks:**
Each magnesium atom loses 2 electrons to form ions with a 2⁺ charge. Chlorine atoms gain 1 electron to form chloride ions. These ions have a 1⁻ charge. The formula for magnesium chloride is MgCl$_2$.

TOTAL MARKS FOR QUESTION 4 = 5

05.1 Advantage: It shows how the electrons are shared **(1)**.
Disadvantage: Suggests the electrons are different when they are really all the same **(1)**.
05.2 Advantage – **any one from:**
- It shows the metal ions are in a lattice
- It explains why magnesium conducts electricity **(1)**
Disadvantage – it does not show that the ions are vibrating **(1)**.
05.3 Advantage – **any one from:**
- It shows which atoms are joined
- It shows the shape of the structure **(1)**
Disadvantage – **any one from:**
- The atoms are too far apart
- There aren't really sticks between the atoms **(1)**.

TOTAL MARKS FOR QUESTION 5 = 6

06.1

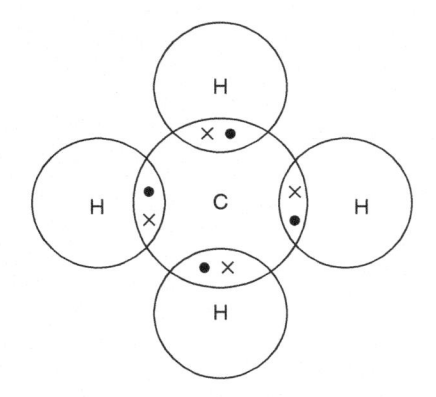

All the dots and crosses must be present to gain 1 mark.

06.2 As the molecules are small, there are weak forces of attraction **(1)** between them **(1)**.

06.3 It has no charged particles that can move **(1)**.

TOTAL MARKS FOR QUESTION 6 = 4

07.1 14 + (4 × 1) × 2 + 32 + (4 × 16) **(1)**
= 132 **(1)**
07.2 2 × 132 = 264 g **(1)**

TOTAL MARKS FOR QUESTION 7 = 3

08.1 X = condenser **(1)**
Y = (round bottom) flask **(1)**
08.2 At X the water vapour condenses to form liquid water **(1)**.
At Y the liquid water evaporates / boils to form water vapour **(1)**.

TOTAL MARKS FOR QUESTION 8 = 4

09 **This is a model answer, which would gain all 6 marks:**
Ria should use sulfuric acid. She should place the acid into a beaker and warm it gently. She should add some copper oxide to the acid and stir. Then she should continue adding copper oxide and stirring until no more copper oxide can react. She will know this has been achieved when she can see unreacted copper oxide collecting at the bottom of the beaker. Ria should filter the mixture using a filter funnel and filter paper, and collect the copper sulfate solution in an evaporating basin. She should gently warm the mixture using a Bunsen burner. The water will evaporate and crystals of copper sulfate will form.

TOTAL MARKS FOR QUESTION 9 = 6

10.1 M$_r$ of magnesium sulfate = 24 + 32 + (4 × 16) = 120 **(1)**
Moles of magnesium sulfate = $\frac{12.0}{120}$ = 0.10 moles **(1)**
M$_r$ of magnesium oxide = 24 + 16 = 40 **(1)**
Mass of magnesium oxide 0.10 × 40 = 4.0 g **(1)**
10.2 $\frac{12}{100}$ × 65 **(1)**
= 7.8 g **(1)**
10.3 **Any one from:**
- Some of the reactants haven't been used up
- The reactants weren't pure
- Some of the product was lost in purification **(1)**.
10.4 120 = M$_r$ of desired products **(1)**
138 = sum of M$_r$ of all the reactants.
$\frac{120}{138}$ × 100 **(1)**
= 87.0 **(1)**
10.5 **Any one from:**
- Less waste
- Less raw materials are required **(1)**

TOTAL MARKS FOR QUESTION 10 = 11

11.1 The electrodes are touching **(1)**.
The electricity would pass from one electrode to the other and not pass through the solution **(1)**.
11.2 It has lost electrons **(1)**.
11.3 Cu²⁺ + 2e⁻ → Cu **(1)**
Copper ions gain electrons **(1)** and form copper atoms **(1)**.
11.4 0.17 **(1)**
11.5 The anode might have been wet **(1)**.
11.6 **Any one from:**
- Repeat the experiment – to minimise the effect of any errors
- Dry the electrodes carefully – as the mass of water can affect the results **(2)**.

11.7

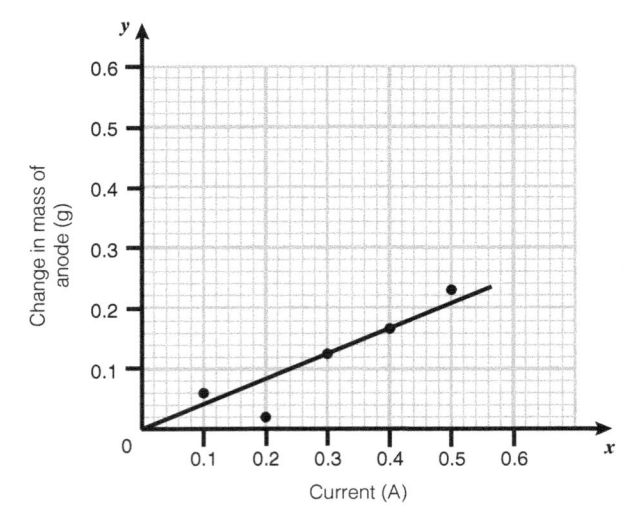

- 1 mark for labelling the x axis
- 1 mark for accurately plotting each point (within half a square)
- 1 mark for the line of best fit **(3)**.

TOTAL MARKS FOR QUESTION 11 = 13

12 Group 1 metals have lower melting points than transition elements **(1)**.
Any two from: Group 1 metals are more reactive / stronger / denser / harder than transition elements **(2)**.
Transition elements form ions with different charges but Group 1 metals only form ions with a 1+ charge **(1)**.
Two marks for using the data in Table 4 to compare properties.

TOTAL MARKS FOR QUESTION 12 = 6

13.1 potassium hydroxide, potassium chloride **(1)**
13.2 Fully ionised / split up **(1)**
Proton donor **(1)**
13.3 $H^+ + OH^- \rightarrow H_2O$ **(1 mark for the reactants and 1 mark for the product) (2)**
13.4 **Any six from (key pieces of equipment in bold):**
- Use a **pipette** to place 25 cm³ acid in a **flask**.
- Place the flask on a **white tile**.
- Add an **indicator** / named indicator to the acid.
- Place the potassium hydroxide in a **burette**.
- Add the potassium hydroxide to the flask – quickly at first…
- … and then slowly.
- Swirl the flask.
- Stop when the indicator changes colour.
- Record the volume of potassium hydroxide added.
- Repeat until you get concordant results **(6)**.

13.5 The results for 2 and 3 should be circled **(1)**.
13.6 $\frac{22.45 + 22.35}{2} = 22.4$ **(1)**
13.7 Amount of acid $= \frac{25.0}{1000} \times 0.20 = 0.005$ mol **(1)**
Amount of potassium hydroxide = 0.005 mol **(1)**
Concentration of the potassium hydroxide $= \frac{0.005 \times 1000}{22.70}$ **(1)**
$= 0.220$ mol / dm³ **(1)**
13.8 $\frac{100}{1000} \times 0.15 = 0.015$ **(1)**
$0.015 \times 56 = 0.84$ g **(1)**

TOTAL MARKS FOR QUESTION 13 = 19

14.1 A shared pair of electrons **(1)**
14.2 The negative electrons in the covalent bond **(1)** attract both the positive nuclei **(1)**.
14.3

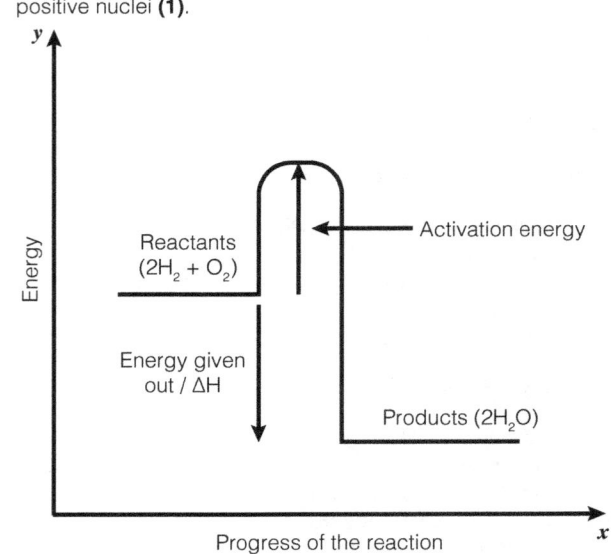

- 1 mark for labelling the reactants
- 1 mark for labelling the products
- 1 mark for labelling the energy given out (ΔH)
- 1 mark for labelling the activation energy **(4)**.

14.4 Energy taken in (when bonds are broken) $(436 \times 2) + 495$
$= 1367$ **(1)**
Energy given out (when bonds are formed) $= (4 \times 463)$
$= 1852$ **(1)**
The energy change $1367 - 1852 = -485$ kJ / mol **(1)**

TOTAL MARKS FOR QUESTION 14 = 10

SET A: PAPER 2

01.1 Heat the nichrome / metal wire in a Bunsen flame **(1)**.
Dip the wire in (concentrated) hydrochloric acid **(1)**.
Dip the wire in the compound and then place it in the Bunsen flame **(1)**.

01.2 Potassium **(1)**

01.3 Sodium ions produce a very intense yellow flame **(1)**.
This masks the more subtle colour from other ions **(1)**.

01.4 Advantage – **any one from:**
- Fast
- Sensitive
- Accurate **(1)**.

Disadvantage – **any one from:**
- Expensive equipment
- Requires specialist training to use **(1)**.

TOTAL MARKS FOR QUESTION 1 = 8

02.1 Carbon particles – Global dimming **(1)**
Nitrogen oxides – Acid rain (also allow global warming) **(1)**

02.2 It contains carbon and hydrogen **(1)** only **(1)**.

02.3 12.5 or 12 ½ **(1)**
8 **(1)**

02.4 **Any one from:**
- Carbon
- Soot
- Carbon monoxide **(1)**.

TOTAL MARKS FOR QUESTION 2 = 7

03.1 It produces smaller, more useful alkanes and reactive alkenes **(1)**.

03.2 C_2H_4 **(1)**

03.3 C_nH_{2n+2} **(1)**

TOTAL MARKS FOR QUESTION 3 = 3

04.1 Same functional group **(1)**

04.2 A **(1)**

04.3 Ethanol **(1)**

04.4 Ethyl ethanoate **(1)**

TOTAL MARKS FOR QUESTION 4 = 4

05.1 Water that is safe to drink **(1)**

05.2 To remove solid particles **(1)**

05.3 To kill harmful bacteria / microorganisms **(1)**

05.4 The water evaporates to form water vapour **(1)**.
The water vapour condenses to form pure liquid water **(1)**.

05.5 It requires large amounts of energy **(1)**.

TOTAL MARKS FOR QUESTION 5 = 6

06.1 Copper hydroxide / copper (II) hydroxide **(1)**

06.2 Blue **(1)**

06.3 $Cu^{2+} + 2OH^- \rightarrow Cu(OH)_2$ **(1)**

TOTAL MARKS FOR QUESTION 6 = 3

07.1 No. They both give the same colour / yellow **(1)**.

07.2 **This is a model answer, which would gain all 5 marks:**
Parminder could place each powder in a separate beaker and add water. Then she would add the silver nitrate solution. A white precipitate would indicate that silver chloride had formed so the powder was sodium chloride. A cream precipitate would indicate that silver bromide had formed so the powder was sodium bromide.

TOTAL MARKS FOR QUESTION 7 = 6

08.1 **Any one from:**
- Objects that have rusted are expensive to replace
- Rusting makes the iron weaker **(1)**.

08.2 Water **(1)** and oxygen **(1)**

TOTAL MARKS FOR QUESTION 8 = 3

09.1 Life cycle assessment / analysis **(1)**

09.2 It helps them to compare products / services **(1)**.

09.3 **Any suitable answer**, such as price; size; cost to run **(1)**

09.4 The top speed of the car **(1)**

TOTAL MARKS FOR QUESTION 9 = 4

10 Environmental impact of paper – **any two from:**
- Paper is made from trees
- More trees can be planted
- Trees are renewable
- Paper is biodegradable **(2)**.

Environmental impact of plastic – **any two from:**
- Plastics are made from oil
- Oil is a non-renewable source
- Plastic is non-biodegradable
- Plastic bags can be used more times
- Plastic bags do not dissolve in water and have to be thrown away **(2)**.

TOTAL MARKS FOR QUESTION 10 = 4

11.1 Propene **(1)**

11.2 Lots of propenes / monomers **(1)** join together by covalent bonds **(1)**.

11.3

$$\left(\begin{array}{cc} H & H \\ | & | \\ -C - C- \\ | & | \\ H & H \end{array} \right)_n$$

1 mark for single bond between the carbon atoms
1 mark for brackets through the trailing bonds and the n **(2)**.

TOTAL MARKS FOR QUESTION 11 = 5

12.1 Ammonia **(1)**; sulfuric acid **(1)**

12.2 The relative amounts of nitrogen, phosphorus and potassium **(1)**

12.3 So they can choose the right fertiliser for their crops / plants / soil **(1)**

TOTAL MARKS FOR QUESTION 12 = 4

13.1 C_2H_4 **(1)**

13.2 Alkenes **(1)**

13.3 **This is a model answer, which would gain all 4 marks:**
The student could place the sample in a test tube and add bromine water. Then they should shake the test tube. If the bromine water decolourises it is ethene. If the bromine water stays the same colour it is ethane.

TOTAL MARKS FOR QUESTION 13 = 6

14.1 A reaction that gives out heat / energy **(1)**

14.2 It is cooled **(1)**.
The ammonia condenses and is piped off **(1)**.

14.3 **Any suitable answer**, such as the gases do not pollute the atmosphere; hydrogen is flammable; less raw materials are needed; it is cheaper **(1)**.

14.4 Iron **(1)**

14.5 Any one from:
- To increase the rate of reaction
- To decrease the time needed for the reaction to reach equilibrium **(1)**.

14.6 **This is a model answer, which would gain all 5 marks:**
A pressure of 200 atmospheres is used in the Haber process. Increasing the pressure favours the side with fewer gas molecules, which is the forwards direction, and gives a good yield of ammonia. The reaction is exothermic. Increasing the temperature favours the reverse reaction but decreasing the temperature decreases the rate of reaction. A temperature of around 450°C is used. This gives a reasonable yield and a reasonable rate of reaction.

TOTAL MARKS FOR QUESTION 14 = 11

15.1 **This is a model answer, which would gain all 5 marks:**
Louise would start by drawing a pencil line near the bottom of the chromatography paper. She would put a dot of ink on the pencil line. Then she would suspend the paper into the solvent in the beaker. Louise should make sure the ink dot is above the water line. She would let the water move up the paper and take the paper out when the water reaches the top of the paper.

15.2 All the components move up the paper / join together at the top **(1)**.

15.3 $\frac{4.5}{16}$ **(1)**
= 0.28125 **(1)**
= 0.281 **(1)**

15.4 There is only one mark **(1)**.

15.5 It is insoluble in water **(1)**.

TOTAL MARKS FOR QUESTION 15 = 11

16.1 HCl, H$_2$ **(for 1 mark)**
2HCl, H$_2$ **(for 2 marks)**

16.2 Any one from:
- Stops the acid spitting out
- Only the hydrogen / gas can escape **(1)**.

16.3

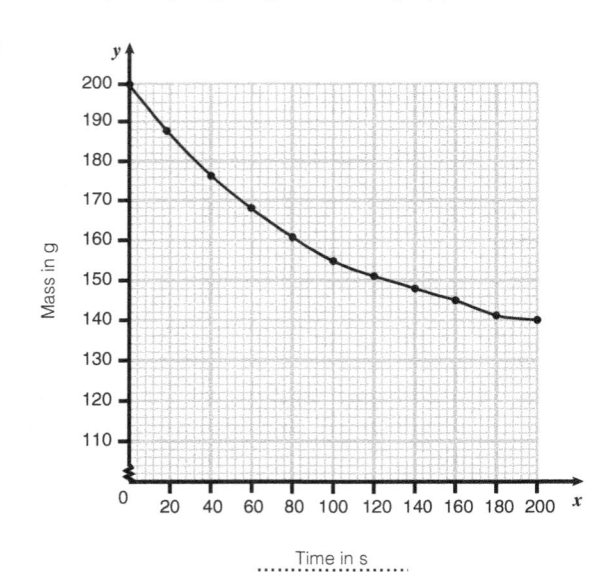

[Graph showing Mass in g vs Time in s, curve decreasing from ~200 to ~140]

- 1 mark for labelling the x axis
- 1 mark for accurate scale for y axis
- 1 mark for accurately plotting each point (within half a square)
- 1 mark for a smooth curve **(4)**

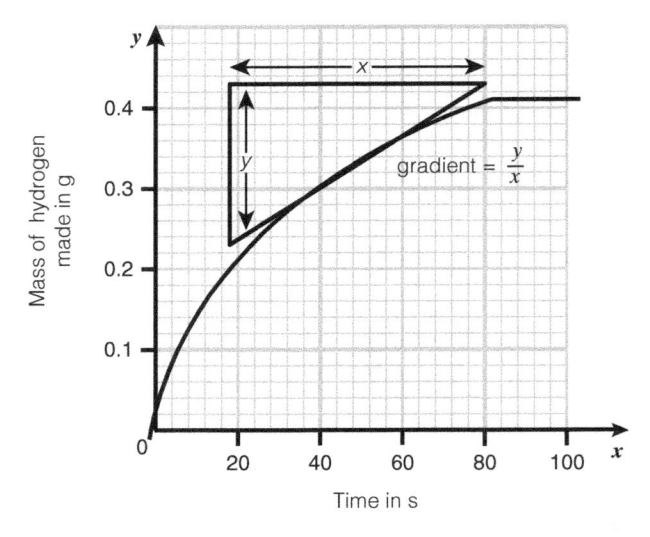

16.4

[Graph showing Mass of hydrogen made in g vs Time in s, with tangent drawn, gradient = $\frac{y}{x}$]

Draw a tangent to the curve at 50 seconds **(1)**.
Calculate the gradient of the tangent (see graph) **(1)**.

16.5 The gradient is steepest at the start = reaction is fastest at the start **(1)**.
The gradient becomes shallower / levels off = reaction slows down / stops **(1)**.

16.6 Particles gain energy and move faster **(1)**.
Therefore, there are more successful collisions in a given time, so the rate increases **(1)**.
Particles collide more often **(1)**.
When they do collide more of the particles have enough energy to react / the activation energy **(1)**.

TOTAL MARKS FOR QUESTION 16 = 15

SET B: PAPER 1

01 Group 1 metals – soft and easily cut with a knife; reactive metals **(2)**
Transition elements – dense; form ions that have different charges **(2)**

TOTAL MARKS FOR QUESTION 1 = 4

02.1 Group 7 **(1)**
02.2 35 **(1)**
02.3 **Any one from:**
- Same atomic number but a different mass number
- Same number of protons but a different number of neutrons
- Atoms of the same element but with different numbers of neutrons **(1)**.

02.4 Bromine-79 and bromine-81 both have 35 protons **(1)**.
Bromine-79 and bromine-81 both have 35 electrons **(1)**.
Bromine-79 has 44 neutrons and bromine-81 has 46 neutrons **(1)**.

TOTAL MARKS FOR QUESTION 2 = 6

03.1 10 **(1)**
03.2 When atoms gain electrons **(1)** or lose electrons **(1)**
03.3 Group 7 **(1)**.
It has seven electrons in its outer shell **(1)**.

TOTAL MARKS FOR QUESTION 3 = 5

04 **This is a model answer, which would gain all 5 marks:**
Each of the two sodium atoms loses 1 electron to form two Na⁺ ions. The compound formed is sodium oxide and has the formula Na₂O. Each oxygen atom gains 2 electrons to form O²⁻ ions.

TOTAL MARKS FOR QUESTION 4 = 5

05 **This is a model answer, which would gain all 6 marks:**
Sodium has a metallic structure / contains metallic bonds. It conducts electricity because the delocalised electrons can move.
Chlorine has a simple molecular structure / contains covalent bonds. It does not conduct electricity because it has no charged particles that are free to move.
Sodium chloride has a giant ionic structure / contains ionic bonds. It does not conduct electricity when solid as the ions cannot move. It does conduct electricity when molten as the ions can move.

TOTAL MARKS FOR QUESTION 5 = 6

06.1 $N_2 + 3H_2 \rightarrow 2NH_3$
(1 mark for correct formula; 1 mark for balancing)

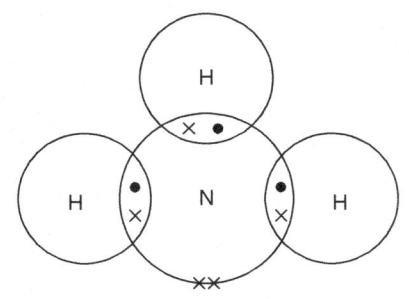

06.2
- 1 mark for bonding electrons
- 1 mark for lone / unbonded pair **(2)**

06.3 As the molecules are small, there are weak forces of attraction **(1)** between them **(1)**.

TOTAL MARKS FOR QUESTION 6 = 6

07.1 Potassium chloride + bromine **(1)**
07.2 Orange **(1)**
07.3 Cl₂ **(1)**.
It has gained electrons **(1)**.
07.4 Hydrogen and chlorine **(1)**

TOTAL MARKS FOR QUESTION 7 = 5

08.1 40 + 16 + 1 + 16 + 1 **(1)**
= 74 **(1)**
08.2 74 × 2.5 = 185 g **(1)**
08.3 $\frac{50}{1000} \times 0.20 = 0.01$ **(1)**
0.01 × 74 = 0.74 g **(1)**

TOTAL MARKS FOR QUESTION 8 = 5

09.1 Metallic bonding **(1)**
Attraction between the positive copper ions **(1)** and the free / delocalised electrons **(1)**
09.2 The metallic bonds are very strong **(1)**.
It takes a lot of energy to overcome them **(1)**.
09.3 The free / delocalised electrons can move **(1)** and carry charge **(1)**.

TOTAL MARKS FOR QUESTION 9 = 7

10 **This is a model answer, which would gain all 6 marks:**
The subatomic particles are protons, neutrons and electrons. Protons have a relative mass of 1 and a relative charge of +1. Neutrons have a relative mass of 1 and no overall charge. Electrons have a negligible mass and a relative charge of –1.

TOTAL MARKS FOR QUESTION 10 = 6

11.1 Circles: carbon atoms **(1)**
Solid lines: covalent bonds **(1)**
11.2 There are lots **(1)** of strong bonds between the carbon atoms in the giant molecular structure **(1)**.
11.3 The free / delocalised electrons / electrons in weak bonds between the layers **(1)** are free to move **(1)**.

TOTAL MARKS FOR QUESTION 11 = 6

12.1 6.02×10^{23} **(1)**
12.2 3.01×10^{23} **(1)**

TOTAL MARKS FOR QUESTION 12 = 2

13.1 **Any one from:**
- There are no forces between the spheres
- All the particles are spheres
- The spheres are solid **(1)**.

13.2 In solids:
- the particles have a regular / uniform structure and are touching
- the particles vibrate **(2)**.

In liquids:
- the particles are touching but do not have a regular / uniform structure
- the particles can move relative to each other **(2)**.

In gases:
- the particles are far apart and have a random arrangement
- the particles move very quickly in all directions **(2)**.

TOTAL MARKS FOR QUESTION 13 = 7

14.1 sodium hydroxide; sodium nitrate **(both answers required for 1 mark)**
14.2 HNO_3 **(1)**
14.3 $H^+ + OH^- \rightarrow H_2O$ **(1 mark for reactants; 1 mark for products)**
14.4 Amount of acid = $\frac{25.0}{1000} \times 0.10 = 0.0025$ mol **(1)**
Amount of potassium hydroxide = $0.0025 \times 2 = 0.005$ mol **(1)**
Concentration of the potassium hydroxide = $\frac{0.005 \times 1000}{28.75}$ **(1)**
= 0.174 mol / dm³ **(1)**

TOTAL MARKS FOR QUESTION 14 = 8

15.1 Covalent bond **(1)**
15.2 Energy required to break bonds = $436 + x$ **(1)**
Energy given out as bonds are made = $(2 \times 432) = 864$ **(1)**
$-185 = 436 + x - 864$ **(1)**
$x = 243$ kJ / mol **(1)**

TOTAL MARKS FOR QUESTION 15 = 5

16.1 It takes in heat energy / the temperature falls **(1)**.
16.2

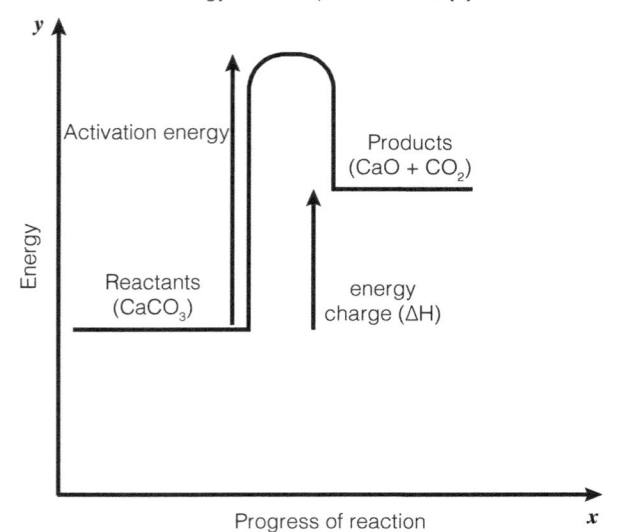

- 1 mark for labelling the reactants
- 1 mark for labelling the products
- 1 mark for labelling the energy given out (ΔH)
- 1 mark for labelling the activation energy **(4)**.

16.3 The minimum energy to start the reaction / break the bonds in the reactants **(1)**.

TOTAL MARKS FOR QUESTION 16 = 6

17.1 M_r of magnesium nitrate = $24 + 2 \times (14 + (16 \times 3))$
= 148 **(1)**
Moles of magnesium nitrate = $\frac{7.4}{148} = 0.05$ mol **(1)**
M_r of magnesium oxide = $24 + 16 = 40$ **(1)**
Mass of magnesium oxide = $0.05 \times 40 = 2.0$ g **(1)**

17.2 $\frac{7.4}{100} \times 70$ **(1)**
= 5.18 g
= 5.2 g **(1)**

17.3 Any one from:
- The reaction does not go to completion
- Some of the product is lost
- Other reactions take place **(1)**.

17.4 166 **(1)**
$\frac{148}{166} \times 100$ **(1)**
= 89.2 **(1)**

17.5 Any one from:
- Sustainable development
- Less waste – waste products do not need to be disposed of **(1)**

TOTAL MARKS FOR QUESTION 17 = 11

SET B: PAPER 2

01 Carbon dioxide: global warming **(1)**
Sulfur oxides: acid rain **(1)**

TOTAL MARKS FOR QUESTION 1 = 2

02.1 When ancient biomass / plankton **(1)** was buried in mud and exposed to high temperatures and pressures **(1)**
02.2 Fractional distillation **(1)**
02.3 Hydrogen **(1)** and carbon **(1)**
02.4 Covalent bonds **(1)**
02.5 $C_9H_{20} + 14O_2 \rightarrow 9CO_2 + 10H_2O$
(1 mark for 14; 1 mark for 9)
02.6 No / not enough oxygen **(1)**
02.7 C_nH_{2n+2} **(1)**
02.8 More flammable **(1)**, so they are better fuels **(1)**
02.9 C_3H_6 **(1)**
02.10 Propene **(1)**
02.11 **Any suitable answer**, such as making polymers; plastics; and alcohols **(1)**.

TOTAL MARKS FOR QUESTION 2 = 15

03.1 Calcium – orange-red **(1)**
Copper – green **(1)**
03.2 Al^{3+}: white **(1)**
Fe^{3+}: brown **(1)**
03.3 Solid **(1)**
03.4 $Ca^{2+} + 2OH^- \rightarrow Ca(OH)_2$
(1 mark for correct formulae; 1 mark for balancing)
03.5 Flame test **(1)**
The one that gives the orange-red colour contains calcium ions **(1)**.
03.6 $Ag^+(aq) + Cl^-(aq) \rightarrow AgCl(s)$
(1 mark for correct formula; 1 mark for state symbols)
03.7 Cl^- – white **(1)**
Br^- – cream **(1)**
I^- – yellow **(1)**
03.8 Hydrogen – Add a burning splint – It burns rapidly with a 'pop' **(1)**
Oxygen – Add a glowing splint – It relights **(1)**
Chlorine – Add damp litmus paper – It is bleached and turns white **(1)**
03.9 Add a dilute acid **(1)**.
Bubble the gas produced through limewater **(1)**.
If the limewater turns milky then carbonate ions are present **(1)**.

TOTAL MARKS FOR QUESTION 3 = 20

04.1 • 1 mark for labelling the energy given out (ΔH)
• 1 mark for labelling the activation energy **(2)**.

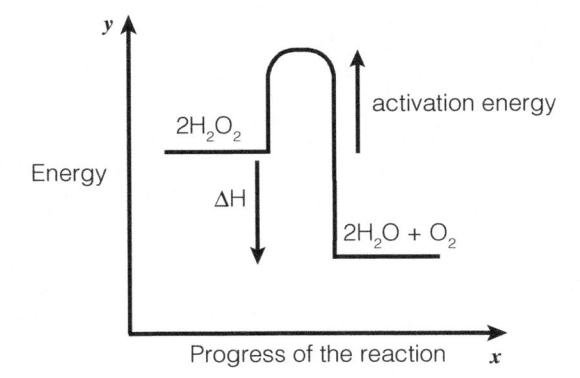

04.2 A line with a lower hump / activation energy and starting at the reactants and ending at the products (see graph) **(1)**
04.3 The catalyst increases the rate of reaction **(1)** by providing an alternative reaction pathway with a lower activation energy **(1)**.
04.4 The catalyst is not used up in the reaction **(1)**.

TOTAL MARKS FOR QUESTION 4 = 6

05.1 OH / hydroxyl **(1)**
05.2 B **(1)**
05.3 $C_3H_7OH + 4½ O_2 \rightarrow 3CO_2 + 4H_2O$
(1 mark for formulae; 1 mark for balancing)
05.4 A: ethanoic acid **(1)**
B: propane **(1)**
C: propene **(1)**
D: propanol **(1)**
05.5 A solution of this compound turns universal indicator orange – A **(1)**.
This hydrocarbon decolourises bromine water – C **(1)**.
This compound forms a neutral solution with water and reacts with sodium to produce hydrogen – D **(1)**.
This compound has the general formula C_nH_{2n+2} – B **(1)**.

TOTAL MARKS FOR QUESTION 5 = 12

06.1 Alloys of gold – Unreactive – Jewellery **(1)**
Stainless steel – Hard and resists corrosion – Cutlery **(1)**
Aluminium alloys – Low density – Aeroplanes **(1)**
06.2 Mixtures that contain at least one metal **(1)**.
Pure metals are too soft / alloys are harder **(1)**.
06.3 Oxygen / O_2 **(1)**
06.4 **This is a model answer, which would gain all 6 marks:**
In test tube A, the iron does not rust as there is no oxygen. Boiling the water removes the oxygen from the water; the layer of oil stops oxygen from the air dissolving into the water.
In test tube B, the iron does rust as both oxygen and water are present.
In test tube C, the iron does not rust as there is no water. The drying agent removes water from the air.
Overall, for rusting to occur, both oxygen and water must be present.
06.5 Magnesium / zinc **(1)**
The more reactive metal reacts **(1)**; the less reactive metal (iron) does not **(1)**.

TOTAL MARKS FOR QUESTION 6 = 15

07 The hydrogen ions are closer together **(1)**.
The hydrogen particles and the magnesium atoms collide more often **(1)**.
More successful collisions mean a faster rate of reaction **(1)**.

TOTAL MARKS FOR QUESTION 7 = 3

08.1 **Any one from:**
• Carbon dioxide
• Methane
• Water vapour **(1)**.
08.2 Carbon capture and storage **(1)**
Tree planting **(1)**
08.3 **Any one from:**
• Scientific arguments about climate change
• People do not want to change their lifestyle
• People do not have enough information
• Some countries may not cooperate **(1)**.

TOTAL MARKS FOR QUESTION 8 = 4

09.1 Ores **(1)**
09.2 **Any one from:**
- It is scarce
- Copper is very useful
- Expensive
- High-grade ores are used up **(1)**.

09.3 Traditional methods involve digging / moving / disposing of large amounts of rock **(1)**.
09.4 Bacteria (produce a leachate that contains the metal) **(1)**.
09.5 Plants absorb the metal ions as they grow **(1)**.
The plants are burnt to produce metal compounds **(1)**.
09.6 Identifies the metal ions present **(1)**
Measures the concentration of the metal ions **(1)**

TOTAL MARKS FOR QUESTION 9 = 8

10.1 Nitrogen is the main gas at about 78% / 80% **(1)**.
Oxygen is the next most abundant at about 21% / 20% **(1)**.
There are small amounts / less than 1% of other gases (carbon dioxide / water vapour / noble gases / argon / neon) **(1)**.
10.2 **This is a model answer, which would gain all 6 marks:**
Intense volcanic activity released gases. Water vapour condensed to form the early oceans. The early atmosphere was mainly carbon dioxide. The early atmosphere had little / no oxygen. The amount of nitrogen increased. There may have been small amounts of ammonia. There may have been small amounts of methane. The amount of carbon dioxide went down. Carbon dioxide dissolved into the oceans forming carbonates. Carbon dioxide was also decreased by the formation of fossil fuels and sedimentary rocks. Algae / plants produced oxygen / increased the amount of oxygen. Algae / plants reduced the amount of carbon dioxide by photosynthesis.

TOTAL MARKS FOR QUESTION 10 = 9

11.1 Poly(ethene); is made in an addition polymerisation reaction **(1)** in which lots of ethene monomers / molecules **(1)** are joined together by covalent bonds as the double bond opens up **(1)**.
11.2 Melamine, as it is a thermosetting polymer **(1)**.
It has cross links between the polymer chains **(1)** so it does not melt when heated / it is much more rigid than poly(ethene) **(1)**.

TOTAL MARKS FOR QUESTION 11 = 6

Notes

Periodic table

Key
- ☐ Metals
- ▨ Non-metals

Key to element box:
- Relative atomic mass → 1
- Atomic symbol → **H**
- Name → hydrogen
- Atomic/proton number → 1

1	2											3	4	5	6	7	0
																	4 **He** helium 2
7 **Li** lithium 3	9 **Be** beryllium 4											11 **B** boron 5	12 **C** carbon 6	14 **N** nitrogen 7	16 **O** oxygen 8	19 **F** fluorine 9	20 **Ne** neon 10
23 **Na** sodium 11	24 **Mg** magnesium 12											27 **Al** aluminium 13	28 **Si** silicon 14	31 **P** phosphorus 15	32 **S** sulfur 16	35.5 **Cl** chlorine 17	40 **Ar** argon 18
39 **K** potassium 19	40 **Ca** calcium 20	45 **Sc** scandium 21	48 **Ti** titanium 22	51 **V** vanadium 23	52 **Cr** chromium 24	55 **Mn** manganese 25	56 **Fe** iron 26	59 **Co** cobalt 27	59 **Ni** nickel 28	63.5 **Cu** copper 29	65 **Zn** zinc 30	70 **Ga** gallium 31	73 **Ge** germanium 32	75 **As** arsenic 33	79 **Se** selenium 34	80 **Br** bromine 35	84 **Kr** krypton 36
85 **Rb** rubidium 37	88 **Sr** strontium 38	89 **Y** yttrium 39	91 **Zr** zirconium 40	93 **Nb** niobium 41	96 **Mo** molybdenum 42	[98] **Tc** technetium 43	101 **Ru** ruthenium 44	103 **Rh** rhodium 45	106 **Pd** palladium 46	108 **Ag** silver 47	112 **Cd** cadmium 48	115 **In** indium 49	119 **Sn** tin 50	122 **Sb** antimony 51	128 **Te** tellurium 52	127 **I** iodine 53	131 **Xe** xenon 54
133 **Cs** caesium 55	137 **Ba** barium 56	139 **La*** lanthanum 57	178 **Hf** hafnium 72	181 **Ta** tantalum 73	184 **W** tungsten 74	186 **Re** rhenium 75	190 **Os** osmium 76	192 **Ir** iridium 77	195 **Pt** platinum 78	197 **Au** gold 79	201 **Hg** mercury 80	204 **Tl** thallium 81	207 **Pb** lead 82	209 **Bi** bismuth 83	[209] **Po** polonium 84	[210] **At** astatine 85	[222] **Rn** radon 86
[223] **Fr** francium 87	[226] **Ra** radium 88	[227] **Ac*** actinium 89	[261] **Rf** rutherfordium 104	[262] **Db** dubnium 105	[266] **Sg** seaborgium 106	[264] **Bh** bohrium 107	[277] **Hs** hassium 108	[268] **Mt** meitnerium 109	[271] **Ds** darmstadtium 110	[272] **Rg** roentgenium 111	[285] **Cn** copernicium 112	[286] **Uut** ununtrium 113	[289] **Fl** flerovium 114	[289] **Uup** ununpentium 115	[293] **Lv** livermorium 116	[294] **Uus** ununseptium 117	[294] **Uuo** ununoctium 118

*The lanthanides (atomic numbers 58–71) and the actinides (atomic numbers 90–103) have been omitted.
The relative atomic masses of copper and chlorine have not been rounded to the nearest whole number.